THE
Riverside Library

KIT CARSON
The Happy Warrior of the Old West

A Biography

By
STANLEY VESTAL

BOSTON AND NEW YORK
HOUGHTON MIFFLIN COMPANY
The Riverside Press Cambridge

COPYRIGHT, 1928, BY WALTER STANLEY CAMPBELL

ALL RIGHTS RESERVED

The Riverside Press
CAMBRIDGE · MASSACHUSETTS
PRINTED IN THE U.S.A.

PREFACE

For nearly thirty years I have felt the need of an authentic portrait of Christopher Carson. Even as a small boy, I knew that something was wrong with the standard biographies, that something was surely missing. And as research mops up the corners and corrects the errors of the earlier accounts of his career, it is more and more clear that the legend needs rechecking.

Kit's first biographers made him out a striking but unaccountable hero. They placed him in a spotlight which threw all the background of his age in shadow, representing him as at once blameless and colorless. The effect was to make the man incredible, and to leave the reader with a hunch that the evidence had been doctored.

To make matters worse, the Western Hero became commercialized, and the country was flooded with showmen, who — for a consideration — posed and postured and made of the Old West a cheap burlesque. This sickening spectacle made us all more skeptical than ever, and Kit Carson seemed about to go the way of the 'noble Red Man' in popular favor. For there was no readable 'Life' to relate the man to the character of the times he lived in, no credible account of the typical product of that heroic age when trick cowboys and professional gunmen were as yet unknown.

It is time to retell the adventures of this great little man. For before long there will be no one left who has talked face to face with his contemporaries, no one left who knows what a tepee smells like, or how a beaver trap was set, no one to make real the background out of which his adventures grew, and against which they must be seen.

For Kit Carson has become a symbol of the American frontier, as Odysseus was of the Greek seafarings, and it is important that we understand and love the thing he represents, that Frontier which made these States a Nation. That is the motive behind this book.

I am familiar with much of the country Kit ranged over, and with that Southwest which he made his lifelong headquarters. I grew up among the Cheyenne and Arapaho Indians, the tribes with which he was most intimately associated, and from which he took his two Indian wives. And I think I have seldom missed an opportunity to talk with an old-timer who could tell me about the days and ways of America's heroic age.

No one can write of the Old West of Kit Carson without a strong sense of obligation to the researches of George Bird Grinnell, Edwin L. Sabin, H. M. Chittenden, R. L. Thwaites, and to Blanche C. Grant, editor of Kit's own brief memoirs, recently published (for the first time) in New Mexico.

In addition to the published sources, both general and specific, which I have consulted, I must acknowledge a great debt to my stepfather, James Robert Campbell, who served on Bancroft's staff and spent much time making investigations in the Southwest; to George Bent, son of Colonel William Bent and Owl Woman, and grandson of the Keeper of the Cheyenne Medicine Arrows, whose recollections of life at Bent's Old Fort and of Kit Carson have been invaluable; and to many of the older Cheyenne and Arapaho Indians of the Southern divisions of those tribes, especially those in and about Colony, Oklahoma: Left Hand, Washee, Watan, Watonga, of the Arapaho: Wolf Chief, Burnt All Over, Roman Nose Thunder, Edmond Guerrier, of the Cheyennes. From all these I have received hints and anecdotes not hitherto published and throwing more light upon the least documented period of Kit Carson's life. I wish also to make due acknowledgment of the kindness of The Macmillan Company, who have generously permitted me to make use of the passages quoted from their 'In the Old West,' by George Frederick Ruxton.

Biographies, like other books, are to be judged by the swiftness and completeness with which the reader is carried into the reimagined world of the past — and by the significance of that world once he is in it. I can only hope that the reader will find here something of that rare epic glamour which has made the materials of this book happy hunting grounds for me.

STANLEY VESTAL

CONTENTS

I. GREENHORN
I. THE RUNAWAY   3
II. THE TRAIL TO SANTA FE   12
III. TAOS AND THE TRAIL   26

II. MOUNTAIN MAN
IV. THE FIRST TRAP   41
V. KIT CARSON, TRAPPER   56
VI. FREE TRAPPER   72
VII. FANDANGO   82
VIII. BLACKFOOT SCRAPE   89
IX. CAPTAIN KIT CARSON   102
X. THE ARAPAHO GIRL   114
XI. THE PRETTIEST FIGHT   129
XII. KIT CARSON'S LUCK   137
XIII. HELL'S FULL OF HIGH SILK HATS   145

III. PLAINSMAN
XIV. THE CARSON MEN   165
XV. THE CHEYENNE WOMAN   175

IV. PATHFINDER
XVI. JOSEFA   189
XVII. THE SPANISH TRAIL   203
XVIII. KLAMATH LAKE   215

V. SOLDIER
XIX. SAN PASQUAL   231

VI. RANCHER
XX. THE LAST TRAP   245

VII. INDIAN AGENT
XXI. Father Kit's Hat   269

VIII. PATRIOT
XXII. The Civil War   275

IX. PEACEMAKER
XXIII. Adobe Walls   281
XXIV. The Last Smoke   294

KIT CARSON

I
GREENHORN

CHAPTER I
THE RUNAWAY

ON Christmas Day in the morning, 1809, an undersized, tow-headed, bandy-legged, blue-eyed boy sped into the world squalling lustily with an uncontrolled excitement which no later adventure could arouse in him. Small, bandy-legged, blue-eyed, and sandy-haired he remained to the end of his days, and to this unimpressive appearance the sun added freckles. Yet this boy, typically backwoods as he was, and apparently no different from other lads of his family and community, was to exhibit such character, display such competence, and achieve such fame as distinguish few other lone adventurers in history.

Dispassionate comparison will demonstrate how worthy he is of a rank even with the best of legendary heroes. Kit Carson's endless journeys through the wilderness make the fabled Mediterranean wanderings of Odysseus seem week-end excursions of a stay-at-home; his humanity rivals Robin Hood's; in readiness to fight and in chivalry to women he rates a *siege* at the Round Table; his courage and coolness against hopeless odds may be matched but not surpassed by the old Norse heroes; while his prowess in innumerable battles — all quite without the aid of invulnerable armor or the encouragement of indulgent goddesses —

makes Achilles look like a wash-out. This is no idle boast; any candid reader will admit it.

Yet Kit was no seeker after renown. Shy and matter-of-fact, he went about the business of his life with no notion that he was to be the archetype of the American pioneer. Before Horace Greeley thought of his celebrated advice, Kit had already gone West and grown up with the country. And because he did grow up with it, he left all the other mountain men behind him — pathetic survivors of a dead epoch. It was this adaptability, this superior competence, which made him the figure he remains in the history of the frontier. When fame came, it abashed him, and he never exhibited any of the showmanship which has so cheapened the Western adventurers of a later day. Kit was no boaster, no outlaw, no charlatan, no gunman. Only the willfulness of youth flung him into that endless series of scrapes, expeditions, sprees, battles, adventures of every sort, making him chief actor on the largest stage whereon an heroic age ever went its swift and roaring way to law and civilization. He looked his part so little that on one occasion some emigrants on the Oregon Trail, having paused to stare at the famous scout, went back to their wagons, hooting and laughing, too smart to be hoaxed by those who had pointed out that insignificant-looking little man.

When fame could no longer be denied, the myth-makers went to work. They piled their legends about Kit until the man himself is hardly seen. They concealed or ignored the wild deeds of his youth, though he killed more men than Billy the Kid; they said nothing of his adventures with women, though he is known to have married three times, and twice without the blessing of the Church. Not knowing how to present such a man, they manufactured a monster. On the one hand they failed to exhibit the winning humanity of their victim; on the other they magnified his exploits, 'laying it on a leetle too thick,' to use Kit's own sly comment on the authorized 'Life.' . . .

The tow-headed boy might have grown up like other small Kentuckians of his generation, one of a family of fourteen children, eating with a long spoon the mush and molasses poured upon the wooden seat of a chair in the log cabin, or sitting on the snake fence in the clearing to munch his hominy and johnny-cake, while he longed for the time when he could handle his own long squirrel rifle and join in the corn huskings and square dances of his elders. But when Kit was a year old, his pappy left Madison County to settle on the Missouri frontier.

The Carsons had been pioneers since the first one left Scotland in the late years of the seventeenth century. Pennsylvania, North Carolina, Kentucky, Missouri — in each they made a home, only to move on as fighting Indians gave way to humdrum farming. They could no more settle down for good than so many sailors. Pappy Carson loaded his family and effects into covered wagons and traveled by slow ox-train with a party of neighbors on the westward trail of Daniel Boone.

On the way through Hardin County they must have passed near the log cabin where Nancy Hanks was nursing the baby boy — Kit's senior by eleven months — who was to preserve the Union which Kit Carson did so much to expand.

In Howard County, Missouri, the Carsons ploughed

their new lands under the eyes of sentries posted at the edges of the fields to guard against raiding Indians. They were content again. The family grew and prospered and Pappy Carson, swerving from the tradition of his tribe, began to plan to make a lawyer of his promising son Kit. The reason for this unusual decision is unknown. Perhaps the example of Andrew Jackson, hero of the Kentucky frontiersmen, swayed the father. Perhaps he thought ten sons one too many to send into the Far West to hunt buffalo and fight Indians. Whatever the cause, the plan came to nothing. Kit was spared the discipline of frontier book larnin'. For when the boy was nine years old, his father was killed by a falling tree.

The boy ran wild with the other children, hunted coons with the houn' dawgs, fought rough-and-tumble in a way that made bigger boys respect his sturdy compact inches, learned to call turkeys and read sign of deer, did chores around the place, rode his brother's ponies, learned to fire his father's heavy rifle, resting the barrel on one rail of the fence. Then he would get buckfever, and miss, but be scairt to cuss the weapon because of what he had heard up at the church-house. And all the time his ears would be full of tales of adventure, of Indian surprises, of Dan'l Boone, of the Revolution and the War of 1812, when at New Orleans, the half-horse, half-alligator men had licked the British after the war was over, just for good measure. It was a good life, that, and proper schooling for the man to be. But it could not last forever. At the age of fifteen, Kit was apprenticed to a saddler, one David Workman, and went to work in his shop in the town of Franklin.

Franklin was then the last town on the fringe of the settlements — the port of embarkation for the great caravans whose white tilts sailed away across the prairie ocean to mountains more distant (in time) than the Himalayas are to-day. The local boosters described their fair city in glowing terms: the Franklin public square contained two acres; the principal streets were eighty-two and a half feet wide; the town afforded 'an agreeable and polished society,' and was 'in business and importance the second town in the territory' — St. Louis, with less than two thousand inhabitants, standing first.

From that square containing two acres radiated the trail up the Missouri River to Oregon and the Northwest and the trail to Santa Fe, the Southwest, and California. When Kit went to work at his bench on one of the principal streets, Franklin was booming with the wild rush of the fur trade — the fur trade, with its tales of vast profits to be made in a season, its desperate adventures, its gay and reckless life. Already Major Long had come to Franklin by the first steamboat, had marched up the Platte to the Rockies and back again by the Arkansas. Every year the Missouri Fur Company studded the Great American Desert with new posts. The Ashley expeditions had vanished into the wilderness and returned with enormous profits and marvelous yarns. The Missouri Legion had joined not uncreditably in the fight with the Rickarees, teaching the Indians that the road to the Northwest must not be blocked. Yellowstone, Oregon, the Columbia, became familiar names. In the Southwest the ventures of Jacob Fowler, McKnight, Becknell, Ezekiel Williams, Bent and St. Vrain had opened the way to Santa Fe.

The Spanish settlements, now free from Spanish tyranny, offered rich markets for all manner of American goods — dazzling prizes for all who dared grasp them.

Up and down the principal streets they passed all day long — great rumbling wagons of Conestoga make with flaring bows soaring fore and aft with the gunwales of the prairie schooner, creaking and tossing among the ruts behind their rolling, indifferent oxen. The steady, slow rhythm of those ox-teams was crossed by the jog and patter of Indian ponies, the smart trot of the carriage horses, the quick scuffle of mules as the herds swung out to the caravans beyond the river in a cloud of dust into which the wrangler vanished, twirling one end of his rope about his head.

Pack trains strung in from the West, from Taos, from El Paso, bringing beaver, bringing buffalo robes, bringing Mexican silver. The packers mingled in the public square and the saloons with keelboat men from up-river, French *voyageurs*, gay and improvident, lank Missouri teamsters in homespuns and checked woolen shirts, rich merchants and fur traders in their high beaver hats and ruffled shirts, self-reliant Delaware hunters, squaws, naked children, and the degraded Indian beggars of the settlements. The newspapers which Kit could not read remain to us and we know what those men talked of: the comings and goings of this trader and that caravan, of riches gained in a year's time, of Spanish markets and Spanish customs officers subdued by a bribe, of mines and mules and Injuns, of buffalo robes and beaver plews in countless thousands.

From the shop Kit could hear the endless bustle of

preparation for the long trail, could hear the fine carriages of St. Louis bankers, dandified traders, pass . . . the cheers and flag-waving and gunshots when a steamboat made the landing . . . the songs of keelboat men and *voyageurs*, the steady march of troops, the lowing of cattle, the wild shouting and carousing of the trappers on their spree.

Those mountain men! How they fascinated the lad in the saddler's shop! Those were the boys who trapped the beaver, fought the Injuns, brought home the bacon, created the wealth in the pockets of the dandies in ruffled shirts. Strong, self-reliant, undisciplined as so many savages, they swaggered into Franklin, throwing their money away with a reckless generosity nothing short of magnificent. What hardships, what hunger, danger, blood, and sweat went into that hatful of dollars which the mountain man gambled away so cheerfully! How he ate and drank, and roared in his cups, boasting of his exploits, condemning the poor greenhorns of the settlements to profound perdition even as they fleeced him.

Day after day the mountain men came to Kit's workbench, brought him their Mexican saddles to repair, brought him bullet pouch and possible sack to be mended, bought his hobbles, lariats, *apishamores*, spurs, or ordered royally whatever struck their fancy for the caparison of horse or saddle mule. They would stand by, looking on at his flying awl, their moccasins shuffling a little on the hard dirt floor, their long fringes of buckskin or Indian hair asway, their old elkskin hunting-coats, black and shiny with the grease and smut of many camp-fires, creased about their broad shoulders, peering out at him from beneath a coonskin cap or old wool hat. They told him yarns of the mountains, of Taos, the trapper town in *Nueva Méjico*, of Indian skirmishes and Spanish *señoritas*, and freely and profanely voiced their scorn of the greenhorns and the settlements and all civilized fixin's whatever.

Kit's head was full of their stories, and all the while his hands were learning what went to the making of a pack-saddle, how a Spanish saddle was rigged, how to cut sweat leathers, saddle strings, *tapaderos*. He learned to fashion the huge sole-leather sheath for the skinning-knife, strongly made with rows of heavy copper rivets, with the queer triangular notch through which the belt passed, the notch which always made the knife slide smoothly, quickly out, no matter how the hand grasped the hilt. He learned how to make and mend gun-covers, fringed and beaded. Before the year was out he was thoroughly familiar with the whole equipment of the trapper. And he turned with distaste to the manufacture of harness for the wagon-masters or the merchants with their fine horses. Headstall, bellyband, reins, crupper! What had a Carson to do with civilization, with silver-mounted harness, with ruffled shirts and claret-colored coats and brass buttons? It was such things, such men that Andrew Jackson had downed in the name of the plain people, in the name of Democracy. What had a Kentuckian to do with smug routine and harness? What had moccasins and homespun to do with silks and fine linen? There was no future, no life, in the saddler's shop.

At the year's end Kit said to himself, imitating the tone and manner of a mountain man, 'Hell's full o' greenhorns! The settlements caint shine with this child, anyways you fix it! Wagh!'

The Runaway

Not long after the "Missouri Intelligencer" carried the following advertisement in its columns:

NOTICE: To whom it may concern: That Christopher Carson, a boy about sixteen years old, small of his age, but thickset, light hair, ran away from the subscriber, living in Franklin, Howard County, Mo., to whom he had been bound to learn the saddler's trade, on or about the first day of September last. He is supposed to have made his way toward the upper part of the state. All persons are notified not to harbor, support, or subsist said boy under penalty of the law. One cent reward will be given to any person who will bring back the said boy.

(signed) DAVID WORKMAN
FRANKLIN, *Oct.* 6, 1826

CHAPTER II
THE TRAIL TO SANTA FE

ON the prairie some distance west of the Franklin ferry, twenty-eight big Pittsburgh wagons belonging to Bent, St. Vrain and Company, Indian and Mexican traders, were being made ready for the long trail to Santa Fe. A few were already ship-shape, their white tilts taut and trim, their tongues — propped on the whiffle-trees — pointing westward, their harnesses in order, fresh tar oozing from the heavy hubs. Others stood gaunt and stripped and empty, with bare, fragile bows spanning the strong wagon-boxes, while all around lay piles of bales and bundles, packs and crates of goods—calicoes and domestic cottons for Santa Fe, beads, paints, stroudings, blankets, firearms and cutlery for the prairie warriors. A hundred men were busy there, heaving up the heavy packages over the end-gates, pushing and pulling them about, stowing them so snugly that no jolting could work them free, no friction damage their contents on that long journey of more than two months.

Here and there a gang of half a dozen men would throw themselves upon the end of a long pole, levering the end of one of the wagons off the ground while the wheels were removed and the axles greased. Elsewhere, rebellious, unbroken mules were kicking and squealing, their stubborn heads drawn up to a wagon wheel with barely an inch of rope to spare, undergoing the starvation preliminary to their first wild experience of harness and the whip.

A group of idlers stood laughing and gesticulating at

a safe distance from a balky Spanish mule, which Blas, the Mexican herder, was vainly trying to lead to its place before one of the loaded wagons. He pulled, he pounded, he grimaced, while vehement Spanish oaths flowed from his bearded lips.

One of the teamsters approached the group, and tapped the shoulder of a muscular, clean-shaven man in a handsome hunting-coat of fringed buckskin. 'Thar's a likely boy hyar lookin' fur ye, Cap'n.'

Charles Bent turned from the amusing contest between Blas and his mule, the smile fading from his jovial, florid face, his long fringes swaying. But the smile returned. Before him he saw a small, stocky youngster in blue jeans and home-made moccasins, tow-headed, freckle-faced, with a huge skinning-knife belted round him, and the muzzle of an old flintlock extending high above his worn coonskin cap. The boy's face was eager, but very solemn.

'I'm the Captain, sonny. What do you want?' The man's voice was friendly, tolerant, good-humored.

The boy's eyes swept him, from the snug Cheyenne moccasins and quilled leggins to the full eye and sleek black hair. He felt he could trust this man, though he did seem something of a dandy.

'I want to go to Santy Fee with ye, Cap'n.' The boy stood very straight, clasping his old flintlock.

'Whose boy are you?'

'Linsey Carson's.'

The name was familiar enough. One of the men volunteered the information that Linsey had gone under. The Captain's eyes traveled down the long rusty rifle from the end of the hickory wiping-stick to the ancient notches on the smooth brown stock.

'Can you shoot? Can you stand guard?'

'Yes, sir.'

'Can you ride?' Idle question this, but Charles Bent liked talking with the boy. He himself had gone up the Platte when he was not much older.

'Yes, sir.'

'Have you got a blanket?' None was in evidence.

'No, sir.' The boy wondered if this meant rejection. Where could *he* get a blanket? 'I reckon I don't need none,' he declared. The men chuckled at his ignorance. Or was it courage?

Charles Bent took his pipe from the embroidered leather case hanging around his neck — the *gage d'amour* of some New Mexican beauty — and began filling the bowl with strong tobacco. The gayety of his lively mother was tempered by the sterner character of his New England sire, the blood of that Silas Bent who led the Boston tea party. He liked the boy's spunk. 'What's your name, son?' No more sonny.

'They call me Kit.'

'All right, Kit. In the morning you go with the train. . . . Tom, give this boy a Mackinaw blanket — and a tin cup,' he added, seeing that the boy had none. . . . The men grinned as Kit shouldered the old flintlock and marched off across the grass to get his traps.

'He'll do to herd the cavvy,' said the Captain.

Having stowed away his equipment under one of the wagons near a mess-fire, Kit Carson passed the balance of the evening watching the preparations for the start. He saw the oxen shod — for a few ox-teams were in use, though Major Riley had not yet demonstrated their powers of sustained endurance on the prairie. He saw the wagon-master direct the stretching of the great Osnaburg sheets over the wagon-bows — two of them for each wagon, with a pair of heavy blankets spread between for better protection — and for the safer smuggling of these contraband articles into *Nueva Mejico*. He watched the haughty Delaware hunters gather round their mess-fire and select their cook for the trip west, and looked with some awe at the group of traders, travelers, invalids from the city gathered round the mess-table before the Captain's tent.

Kit himself fed with the teamsters and the mountain men on fresh bread, salt sow-belly, hot coffee from a tin cup big as a pint bowl. Silent and wary, he sat within earshot of the mountain men about their fire, listening to their tales of Injun scrapes, learning how so-and-so got rubbed out near Pawnee Rock; how such-another lost his animals and had to *cache* his packs in the Black Hills; how the Bent brothers were talking of building a new fort on Arkansas, 'count of them Cheyennes and 'Rapahoes movin' south to steal horses from the Kioways and Comanches on the Staked Plains.

Every one was lively that warm August evening, every one happy to be on the way. The trappers were broke and in debt again for new traps and ammunition, sick of civilized doin's, grease hungry for fat cow and antelope steak. One poor fellow, pale and shaking, with blue lips and bloodshot eyes, sat on a wagon tongue, victim of the horrors after his week's debauch, and even then, maybeso, seein' snakes and sich.

But no poverty, no hardships could quench the spirits of the French Canadians. Merry and happy-go-lucky as usual, they roused the caravan with their gay singing — glad merely to be alive and full-fed and in the company of their kind. Careless of the labor of the coming day, they sat and sang most of the night.

Kit was long in going to sleep, though he had made his bed in exact imitation of the mountain men around him. The heavy blanket seemed very thin, and the prairie pushed hard against his unseasoned bones. He would not feel quite easy until the wagon train had moved away from the river, safely out of sight of Franklin, away on the boundless plains where no law could seize a runaway apprentice and drag him back to labor at the saddler's bench on that square containing two acres.

The sky was still dark when the wagon-master roused the sleeping men with his loud 'Turn out!' They got up slowly that first morning, rolled and tied their blankets and tossed them into the wagons. That was hardly accomplished when they heard the second command, 'Catch up!' Then they harnessed the teams with much swearing and the squealing and kicking of half-broken mules. The sullen bulls, new to the yoke, dawdled heavily in spite of the sharp goads. But at last one teamster, quicker or more fortunate than the rest, sang out, 'All set!' One after another took up the cry, as each one got his team in order; and when all had reported, the wagon-master gave the command to march: 'Stretch out!'

It was an impressive sight to the greenhorn Kit, the way those twenty-eight huge wagons, each drawn by an eight-mule team, streamed out to the west in two long files. Besides the wagons there were the carriages and horsemen of the Captain and his mess, the ponies of the Delaware hunters, the mountain men's saddle-mules and pack-animals, the teamsters marching beside their swaying oxen.

Last of all, scrambling along behind through the dust and the ruts, a miscellaneous herd of spare animals followed: lame oxen, extra saddle-horses, sore-backed mules, untamed ponies, mares with foals trotting blindly about — all excited, unstable, likely at any moment to stampede back to the settlements, back to their familiar pastures. This herd of spare animals, this mob of uncongenial beasts, this cavvy it was Kit's duty to wrangle as best he could. He followed them swinging one end of a rope, heading back the strays, whipping up the laggards, learning in ever-repeated lessons the cussedness of mules, the waywardness of horseflesh, the unshaken dumbness of oxen. He was mounted on a slow animal, for he was no hand at the rope as yet, and had had to ride the first — and stodgiest — horse he could catch.

For two weeks the caravan plodded along, halting during the heat of noon for breakfast, making a miserable ten or twelve miles a day. Then Council Grove was reached, and the men and animals seemed to settle down to the business of the long trail; for beyond the Grove was the real plains country, plains growing ever more arid, wilder, more dangerous, where Indians and buffalo might be encountered.

In those two weeks, Kit had learned how to find strayed animals, how to handle a rope with some hope of success in noosing a dodging mule, how to sleep soundly in spite of hard ground, mosquitoes, stamping mules, talk and singing. He had learned that a horse makes for water after feeding — a bit of knowledge that saved him many a long tramp, enabling him to find the strays by going to the nearest stream. He had found that the best place to rope an animal was the corral formed by the wagons, and that it was wise to be first up in order to do it before the wagons stretched out on the trail. He learned to go without food until noon, and to endure the soaking of a thunder-storm without complaint, though he had no tent, and the wagons were too full of precious goods to admit even so small a person as he was. And he learned to sit and listen — to the mountain men, to the Indian hunters, to the teamsters, even to the greenhorn sportsmen with their fancy clothes and scatter-guns.

And now he had to stand his tour of guard duty, four hours every third night, standing in the darkness, whatever the weather, just beyond the ring of tethered mules outside the corralled wagons, flintlock loaded with ball, eyes and ears alert for the first sign of Injuns. There is a story that he shot his own mule one stormy night, mistaking it for a marauding Pawnee. But the story has little credit with the judicious. Kit at this time had no mule to shoot.

As the caravan approached the buffalo range, wolves became numerous. White wolves, especially, abounded, and the mountain men declared that the herds were certainly not far ahead. The teamsters, happy at last in having a mark for their useless rifles, delighted in shooting at the lazy, skulking beasts. One of these men, named Broadus, in his haste to bring down a particularly large wolf, which was making off over the prairie, ran to the tail of his wagon and began to pull his gun out by the muzzle as he ran. Somehow or other, the gun went off, and the ball shattered the bones of his forearm to bits.

The train was halted, and people began asking if there was a surgeon in the caravan. There was not.

But Charles Bent and the wagon-master urged the injured man to let them amputate, telling him that otherwise it was only a question of days until he would be a dead man. But the man with the broken arm would not consent. His nerve was gone, and he refused to permit green hands to experiment on him. So they made room for him in one of the carriages, and the caravan moved on. They all regarded him as a dead man.

Next day they reached the Grand Arkansas and about noon heard a confused, dull murmuring sound, which seemed to come from a distance, and grew louder and louder as they advanced. Soon after, dark masses showed on the plains ahead, and cries of '*Bison! Cibola! Buffalo!*' rang out along the train.

Kit strained his eyes through the dust of the cavvy, and his heart beat fast. Everywhere the hunters and sportsmen were mounting and riding off, everywhere there was bustle, shouting, rejoicing. Now there would be a relief from sow-belly and dry bread; now there would be fat cow and *boudins*, hump-ribs and tongue. Now Kit would taste his first deep-red buffalo meat.

The caravan moved steadily on, passed a few old bulls rolling in the dust to windward, saw the herds ahead all in motion athwart the trail, saw white puffs of smoke as the hunters neared the dark masses, saw them roll up the dust and vanish, hunters and all, in that dun cloud. How the teamsters longed to join the sport! How Kit would have loved to ride his pony after those flying herds, to run meat in genuine trapper style, to bring down fat cows for the mess!

But the cavvy claimed all his care just then, even with Blas and his herders coming to help. The mules were wild with excitement, eager to join the galloping buffalo, and Blas knew well enough that once a mule ran into the herds, he was gone for good. The chances of recapture were negligible. So Kit remained to help with the cavvy. His father's old flintlock had no opportunity to throw buffalo that day.

Before they halted for the night, the train found the Delaware hunters waiting by the trail, their ponies loaded with the choicest cuts. That night around the mess-fires there was gayety and good living, and every one ate until he felt like bursting. Kit heartily agreed with an old trapper, who declared, 'Civilized doin's caint shine with fat buffler, anyways you fix it.'

On Walnut Creek they halted for a day. Broadus was dying, they agreed. His arm was so gangrened that spots had appeared well above the place where the amputation should have been performed. The poor fellow, suffering agonies as he rode over the rough prairie roads, and hoping against hope, now began to plead with the men to perform the operation which he had refused to permit before. It seemed no use to them. It would only kill him outright, and they wanted no hand in his death. But he was so urgent, so persistent, that at last volunteers were called for.

Broadus lay on the buffalo grass, sick with terror and pain, begging the men not to let him die, arguing, crying, pleading, while they looked on, stirring uneasily, glancing at each other, chafing at their inability to help, curious, pitiful, nervous. They all looked at the wagon-master, who waited for an answer to his call.

Kit Carson, moved by the man's agony, spoke up. 'I kin do it.' The men stared at him, and commented, 'Yo're too young, Kit.' ... 'Let somebody else do it.'

... 'Shore, I'll help.' ... Following the boy's lead, shamed out of their indifference, others volunteered. After that, there was no delay. A skinning-knife was whetted to razor sharpness. One of the teamsters brought out an old rusty handsaw from his tool-box, and the back of this was filed to a set of fine teeth. A small fire was built, and the king-bolt of one of the wagons laid upon the coals. When all was ready, they placed the patient on his back on the grass, and a dozen men held him fast while the amateur surgeons went about their terrible deed of mercy. The tourniquet prevented bleeding. The whetted knife quickly opened the arm to the bone; the bone was immediately sawed off; the white-hot bolt seared the raw stump, taking up the arteries more swiftly than ligatures could have done, had they had any. Then a coating of cool axle-grease was laid over the wound and covered with improvised bandages. The patient was carried to his bed in the shade of one of the wagons. Next day the caravan moved on.

What part Kit played in this remains uncertain. But the incident marked his memory as with a hot iron. In his account of the trip to Santa Fe it occupies an altogether disproportionate space. Then for the first time the savage cruelty of life in the wilderness was brought home to the greenhorn boy, while he stood by, silent, pale, listening to the screams and prayers of the suffering teamster, doing his little to help. From that day the quick impulsiveness natural to him was tempered by a caution which nicely balanced the swift decision and passionate action so characteristic of the man. Kit never forgot the sights and sounds and odors of that hot afternoon beside the sprawling Arkansas.

The caravan passed Pawnee Rock, then a bold sandstone promontory jutting toward the river. There the prairie travelers were accustomed to carve their names and the date of their passage. Kit could neither read nor carve the letters of his name, and so Pawnee Rock lacked interest for him. He could not foresee that long afterward this landmark was to be associated with an apocryphal legend — a legend in which he played the part of a hero. To him it was only a landmark, three hundred miles from the settlements, almost halfway to Santa Fe.

Steadily the caravan advanced, following the windings of the river — low barren banks, unmarked by trees; bare sandhills away beyond the grassy flats; shallow interlacing rivulets, vainly attempting to lay claim to the whole of the broad sandy bed. And all around, far as the eye could reach, the brown parched prairie grass, which crunched like snow under the moccasins. And when the caravan halted on the treeless plain, Kit would take his saddle-blanket and in it gather those round dry buffalo chips which served in lieu of wood for the fires.

The Bent caravan did not follow the direct route to the Spanish settlements across the Cimarron Desert, but kept on up the Arkansas to the new stockade belonging to the firm beyond Fountain Creek toward the mountains. The wagons containing goods for the Santa Fe trade diverged to the south, crossing the river above the Big Timbers, either near the mouth of the River of Lost Souls or Huerfano Creek, or perhaps at the mouth of Fountain Creek itself. Charles Bent went on to Santa Fe. His partners, William Bent and Ceran St. Vrain, met the train at the forking of the trail, and it was there that Kit Carson made acquaintance with William Bent.

Passing the Big Timbers, the caravan encountered heavy storms. In the afternoon great black clouds swam up out of the west, swept over the whole sky, passing overhead with the speed of a race-horse, showing the color of cut lead. Then came the dust-storm, followed immediately by a deluge, as though a lake had fallen bodily from the sky. As it happened, the wagons were corralled for the midday halt, and Kit was free to shelter himself, leaving the cavvy within the pen formed by the wagons. Now that the settlements were far behind, the animals showed little disposition to run away. When Kit saw the storm coming, he made a dash for the trees along the river and stumbled upon an old deserted fort of logs thrown together, memorial of some forgotten Indian skirmish. Alongside was a war-lodge of poles covered with boughs, the quarters of the warriors who had built the fort.

The storm was close upon him, and Kit ducked into the war-lodge just in time to escape a drenching. He remained there, lying among the dried grass of the Indian beds, until the storm was over, and went back to the soaked mess feeling very cocky because he was warm and dry. But by the time the caravan reached the fork in the trail, he was miserable, sore, scratching every moment. The graybacks left by the Indians had adopted him.

At the last camp on the Arkansas, Kit sat on a wagon-tongue scratching himself as usual. He looked up, and saw a tall, wiry, fierce-looking boy of about his own age, with black hair and eyes, wearing buckskins.

As their eyes met, William Bent grinned. 'What ails ye?'

'I reckon you kin see,' said Kit, grinning back; 'graybacks.'

'Been sleepin' in some Injun wickiup, I'll lay.'

'Yep. I'm lousy.'

'Want to get rid of 'em?'

'I sure do.'

'Come along with me.' William led the way out on the prairie, his eye searching the brown expanse of parched grass. 'Here you are. Take off your clothes, turn 'em inside out, and throw 'em yonder.' He pointed to a bare circle of earth covered with coarse grains of gravel. Huge red ants ran to and fro over their mounded nest. 'Then you go and soak your hide in the river for an hour or so. The ants will soon clean out all those varmints.'

Kit grinned. Bill was smart. He was some punkins, sure enough. Hastily he stripped and tossed his shirt of fire upon the broad bare circle in the grass. Then he went off to soak in the river. Bill Bent followed, telling Kit of his two years' experience as a trapper, of the new stockade they were building up-river, of the fortunes they hoped to make in the Indian trade. It was the beginning of a lifelong friendship.

That night Kit slept well for the first time in days. Next morning the Santa Fe wagons moved southward. We have no details of the journey. But finally the cluster of angular, flat, adobe houses was seen from the brow of the hill, and the teamsters, their hair slicked with bear's grease, their long whips tipped with new lashes to crack as they passed the staring populace, cheered with excitement. The long string of wagons and carriages and horsemen rolled merrily down the hill, through the narrow streets, into the big plaza.

Kit Carson sat on his mule, his quirt dangling, his tired body propped on the horn of the saddle, his curious eyes ranging over the pillars of the Governor's Palace, the façade of the Cathedral, the shop fronts and mud walls of the houses. So *this* was Santy Fee?

CHAPTER III
TAOS AND THE TRAIL

KIT CARSON did not linger in Santa Fe. The coming of the caravan was celebrated, as usual, by all manner of festivities, but in these the boy had little part. For a day or two he hung around the plaza with the men of the train, exploring such narrow lanes as Burro Alley, watching the teamsters gamble away their hard-earned cash at *monte* in the saloons, standing wedged in the ring of spectators at the cock-fight, staring in at the dark-faced *señoras* who capered to the music of tom-tom and guitar at the fandangos. Then he said good-bye to Broadus and the rest, and accepting the Mackinaw blanket and tin cup and a handful of Mexican silver as his pay for driving the cavvy those eight hundred miles, he joined a party of mountain men and headed north for Taos.

He had expected to walk, carrying his plunder on his back and the long, rusty rifle on his shoulder. But the old trapper Kincade, who had made friends with Kit on the long trail out, had a spare saddle horse he let the boy ride. Kit was glad to go. It was November, and the mountain winter was about to set in. Santa Fe, with its few thousand people, offered no work for a greenhorn boy. The town was full of *gringos* — strong men — men who could speak Spanish. Kit might have starved in Santa Fe. But he never dreamed of returning to Missouri. Taos would be hospitable. It was more friendly to Americans. Best of all, for a boy of Kit's ambitions, it was trapper headquarters.

The trail to Taos was not long, only eighty miles or

so, and even in 1826 had been well traveled by white men for nearly three hundred years and for countless centuries before that by the Indians. Santa Fe, Taos, who can guess how ancient those towns were, even before the *conquistadores* came looking for gold and slaves and converts to the Holy Faith? That trail seemed to pass through a wilderness, but it was a road of long-established life, of a culture far older than that of the upstart colonies along the Atlantic from which Kit and his fathers had derived.

The party left Santa Fe, passing among those strangely depraved hills studded with stunted pinyon, hills pink and salmon in color, dead hills or dead-alive, looking like hams stuck with cloves. The mountain men pushed on, over those hills, up the valley of the Rio Grande del Norte, crossing those broad, dry, cobbled *arroyos*, those rain roads (as the Indians called them), where torrents rushed down to the river when the cloudbursts fell. Past Santa Clara Pueblo they went, past Santa Cruz, through the pass at Embudo, up the canyon and out of it, until they could see, across the sandhills and the tufted sage, the whitewashed walls of San Fernandez de Taos on its hill, shining in the afternoon sun.

A short ride brought them clattering into the Mexican village, and they halted in the small plaza before a shady *portal*, where the shadows of the projecting roof-beams fell in sharp diagonals across the dazzling whiteness of the wall. They got off their mounts a little stiffly after the long day's ride. From the dark door — so thick-walled it was more like a tunnel than a door — came the proprietor of the American House, hand outstretched in greeting.

'How are ye, Kincade, old hoss? I swar ye look done-up. Come in and take a horn of Taos Lightnin'. It'll do yore old guts good.'

'*Wal*, it will. Jest yank the corncob outen the jug, and I'll swaller it as easy as lickin' a dish. . . . And cook us some fust-rate doin's. Me an' this boy hyar aint et fur so long we don't heft as much as our own shadders.'

'Come on in, then. I reckon thar's a chance of potaters on the place,' said their host, and led the way into the warm gloom of his mud-walled establishment.

Kit and Kincade sat at table for the first time in months — not without some grumbling on the part of the old trapper, whose rheumatism made any change of position a trial. There Kit sat hungrily, listening in silence to the gabble about Santy Fee, St. Louie, Injuns and the price of beaver, and guardedly looked round the room, taking in its simple furnishings. . . . But at last they were feeding on beef and potatoes and hot coffee, to which Kincade added sundry potations from the black bottle of *aguardiente* which Ewing Young had set before him. The quiet room, the long bar, the rows of kegs and bottles, the smell of food, the table, the fire, and the friendly welcome warmed the heart of the boy, Kit Carson. These were men of the sort he understood, admired, envied, aimed to copy: after all the hardships of the trail, here was shelter, fire, food, rest, and friends. Taos seemed like home that day, and from that day to the end of his life Taos was home — a home he seldom had the leisure to visit.

That winter the boy had plenty of time to get used to the Mexican village. It does not take long for an active, inquiring boy to make friends with five hundred people, and in winter Taos had rather less than that number of inhabitants. As Kincade said, you could kiver the hull place with a saddle-blanket.

Kit liked it all. The whitewashed, crumbling, flat-roofed, one-story *adobe* houses, with their mica windows guarded by iron bars or painted wooden shutters. The narrow, unpaved, winding lanes and alleys leading to the plaza and its naked cottonwoods. The long, shadowy, echoing *portales*, the rigid *vigas* thrusting their ends out through the walls, the heavy sagging gates of the hidden *patios*. The irrigated gardens, the fields, and the twin red-brown communal houses of the Indian pueblo up the stream. The soft gloom of the quiet interiors with their queer modeled fireplaces, their strings of red peppers, their gay Indian rugs stacked along the walls for seats by day and beds by night, their Pueblo pottery and baskets, their silver-mounted saddles hung on pegs, their *santos* and grotesque holy pictures. Outside, their chimneys made of broken pots, their beehive ovens, and the white, white bread that was baked in them. Mountain, and meadow, and 'dobe wall, Kit liked them all.

In the streets prowling dogs, cats perched on the house-tops sunning themselves, burros drooping lank, despairing ears over the refuse of the alleys, herds of sure-footed, cynical goats pattering through the town, sheep and shepherds, *rancheros* driving strings of tiny, mouse-gray donkeys down from the mountain, their patient heads outlined against the enormous halo of neatly cut firewood piled incredibly upon their backs. Oxen dragging heavy *carros* with solid wooden wheels — ungreased — that groaned and squeaked and shrieked under their loads. Holy processions from the big church, images in tinsel painted to grotesquerie and carried tossing above the heads of the kneeling people to the music of tom-tom, *bandolin*, guitar, amid the clank of bells from the crumbling towers.

And the people. Indians from the pueblo near by wearing costumes which told plainly of their position midway between the Desert and the Plains. Spaniards — as the Americans persisted in calling the native Mexicans — took some getting used to. How Kit stared at first at the fiercely mustached men in their heavy glazed sombreros, gay *sarapes* draped round their shoulders, colored shirts, leather pantaloons unbuttoned halfway up the leg to show the white drawers within. Kit was a little abashed by the exotic, black-eyed girls in their short skirts, skimpy white chemises, their bare shoulders half hidden beneath gay *rebosos* or sober black *mantillas*. He found their smoking of cornshuck cigarettes a more graceful gesture than the sucking of a corncob pipe in the mouth of a Kentucky granny. He wished he understood the glib Spanish they rattled off amid their laughter as he passed. He did not understand then how favorably they compared the sturdy, self-reliant *gringo* youth with the envious, polite *pelados* of their own breed. Rude as the mountain man usually was, and scornful of the courtesy so inbred in the darned Spaniards, he was nevertheless a great favorite of the Taos girls, not to say wives. And this even though the women could not regard him as a *Cristiano*, much less a *caballero*. Foreign women notoriously make excellent teachers of their languages, particularly when the pupils are young men. And although New Mexican convention did not contemplate conversation between unmarried men and women, the mountain men never stood upon convention in these matters.

Whoever was his teacher, one thing is certain. Kit Carson learned to speak Spanish that first winter in Taos. Then, as always, Taos was the resort of many transients, a center of trade and barter between the Indians of the mountains and the Indians of the Plains, a rendezvous for traders from Santa Fe and the States, the Arkansas and the Platte, an outpost of the Mexican customs and the Mexican army, a haven for invalids, a place for outfitting hunters, explorers, miners, prospectors, English sportsmen. And from Taos set out those brigades of trappers for the mountains and the desert to return, empty-handed or rich in furs, for the wasting of their substance in riotous living. Great fairs were held there, great dances and ceremonies at the Indian pueblo near by. There was gambling, dancing, bull-tailing and cock-fighting on Sundays and feast days. And there was always the spice of danger — always the chance that Navajo or Apache would come riding, raiding the village as they had raided the Spanish settlements year after year time out of mind, as often as the season sent them forth to plunder.

Old Kincade had had to give up trapping. His bones had been through too many icy beaver streams, and the rheumatiz had him bested. And so the graybeard sat in his mud-roofed hovel in Taos that winter, telling tall tales of his prowess to Kit, drinking raw Taos whiskey, teaching the boy how a gun was repaired. For old Kincade got a few dollars from the trappers for repairing their weapons, running pig-lead into bullets for them, selling them Galena pills and Du Pont powder with which to run meat or fight Injuns. On the side he helped in the great American game of smuggling then going on, and had a small share in the distillery outside the village where the trappers' fiery liquor was manufactured.

In the spring Kit drifted down to Santa Fe, broke and desperate for work. Old Kincade had gone under, and Kit had to fend for himself. In Santa Fe he was lucky enough to get a place as teamster with the first caravan bound for the States. The job saved him from hunger, but he did not think that luck. He hated going back. But he was so hungry that his belt-buckle rasped his backbone, and he had to go.

The wagon-train took the route across the Cimarron Desert. Up the hill from the plaza, past scattered *ranchos*, past the last wretched Mexican settlement on the Pecos, on beyond that on Turkey River where a single hovel stood. The country grew ever more open, more level, and the mountains receded into hazy blue mounds far to the west behind them. Buffalo appeared, and for the first time Kit encountered the Spanish buffalo hunters, *ciboleros*. They were picturesquely mounted upon small ponies, and clad in leather trousers and jackets, and flat straw hats. Each carried on his back a quiver for his bow and arrows, and from the pommel of his saddle hung on one side his lance decorated with tassels of gay, parti-colored wool, and on the other an enormous Nor'west fusil, with a tasseled wooden stopper in the muzzle. Having exchanged the news of Santa Fe for that of the Plains ahead, the caravan moved slowly on, passing south of the Rabbit Ear Mounds to encamp one night on the Cimarron. There they filled every keg and bucket with water, for the dreaded Desert of the Cimarron lay just ahead — a two days' dry march or water scrape. The cooks prepared food for two days, the animals rested, the men talked of the dangers of losing one's self in that region so lacking in landmarks, so thirsty, with so hard a surface that no amount of travel would leave a trail that could be followed. The wagon-master laid his course by compass.

Progress was slow, the water became exhausted, and the weather was desperately hot, so that men and animals suffered extremely. At last they reached the ford of the Arkansas, the Cimarron Crossing as it had come to be known, and by hard labor snaked their wagons over the quicksands to the grassy flats. From the Cimarron Crossing to Franklin was plain sailing. Kit recognized the place as part of the trail he had followed out, and Missouri seemed suddenly very near. He felt gone beaver that night. It looked to him as if he was about to put his foot in the biggest kind of trap.

Next morning, while his caravan rested, another hove in sight from the east, and the men of the two trains made haste to fraternize, to gather the latest news, to trade and gossip. Kit learned that the westbound train was short-handed. That was enough. He offered his services, took leave of the caravan bound for the settlements, and transferred his plunder to the wagon of his new employer. That night Kit was happy. The dreaded Cimarron Desert was just ahead. But he was going back — back to Taos — to the mountains — to the land of Injuns, beaver, trappers. Surely, sooner or later, he would have a chance to join a brigade of mountain men. If he could only hang on — hang on!

Back he went, found himself in Santa Fe in the autumn, broke again. He had lived through the summer, that was all. But now he needed more than food. He needed something absolutely essential — a woolen shirt. He could dress buckskins and make himself other garments, if need be; but a woolen shirt he had to buy. It was chilly, and winter was bitter in the mountains. Kit parted with his skinning-knife — with everything but the old rusty flintlock. Even that was on its way to be hocked when he heard of a man who wanted teamsters to drive to El Paso. Kit went to work again. The flintlock was saved.

All winter Kit drove team, then took his pay and made tracks to Taos again. Taos was home. He had not run away and worked all these months to become a teamster in the Spanish settlements. So he turned up in Santa Fe again, found no work there, went on to Taos. Arrived — broke as usual.

'Sure,' said Ewing Young, 'I'll give you a job. Can ye cook?'

'Never tried it yit,' Kit confessed.

'Wal, ye'll never larn any younger. You're hired.'

'What do I git outen it?'

'Show me yore cookin' fust.'

Kit went into the kitchen, got together a meal of potatoes, buffalo meat, hot coffee, flapjacks. He brought it in and put it on the table before Ewing Young. Young looked at the mess. 'Set down, Kit, and help yerself. Seems like I kinder lost my appetite to-day, somehow.'

' 'Pears like you think I'm a pore make-out of a cook. But you hired me. Don't forget that.' Kit tried to eat what he had brought in. Tried hard. Failed, hungry as he was.

'Sure, I hired ye. I'll stick by my word, Kit.'

'What do I git outen it?'

Ewing Young looked over his beard at the stocky

youngster with the cold blue eyes and steady hands. There was the making of a man there. He sighed to think of the food he would have to eat until Kit learned how. 'I reckon I'll hev to give ye yore board, Kit,' he said, grinning. 'Yo're stout; maybeso ye kin stand it.' 'I kin if you kin,' said Kit.

Kit Carson's training in cookery was not hindered by any lack of advice or criticism. The mountain men and traders, the hunters and plainsmen who frequented Ewing Young's house were frank, to say the least. Meat was the staple food, and of meat those men professed to be connoisseurs. Buffalo, venison, antelope, mountain mutton, turkey, grouse, water-fowl, jackrabbits, cottontails, bear, and even such supremely delicious flesh as that of panthers, Indian dogs, and beaver tails — they had tasted and prepared them all. Their criticisms made Kit wince behind his grin, sometimes. But he took to heart their well-meant advice, and by the end of winter could say to his employer with a certain pride, 'I reckon I got the best of the bargain after all.'

And Ewing Young replied, as he chewed the rich red buffalo steak Kit had placed before him, 'I dunno, Kit. I swar even painter meat caint shine with this hyar.' And he wagged his shaggy head in confirmation.

That spring Kit went down to Santa Fe again. Wanderlust was upon him, and cooking was getting him nowhere, even though he was fat and sassy as a wolf pup. In Santa Fe he signed up with a caravan going to Missouri and set out once more on the long trail.

The old route was now abandoned, and a more direct road was followed, somewhat to the north of the last year's line of march. The Santa Fe trade was fast settling into a well-organized business, and the halts on the new route were to be remembered as marking the Old Santa Fe Trail. To the men who drove up and down that trail year after year, its landmarks, its camping places, became as familiar as an old shoe. They knew that road from end to end, as a man knows the street before his house.

San Miguel, Rio Mora, Ocaté, Point of Rocks, Rabbit Ear Mounds, the Cimarron water scrape — the train passed them all. There were stories of Indian outrages now, of how the Arapahoes had visited a small party of traders, had intimidated them into giving up thirty horses — one for each warrior in the group. Finding this so easy, they then demanded two horses each, and ended by running off the entire herd of the unlucky white men, who had to walk, leaving their goods to rot on the prairie. And all this after the traders had entertained and fed them!

Said Big Jim Lawrence, mountain man: 'Them 'Rapahoes are the most unsartainest critters in all creation. I reckon they aint more'n half human, if it comes to that. You never heerd of a human, after you'd fed him and treated him to the best doin's in yore lodge, what'd jest turn around and steal all yore hosses and everything else he could lay hands on, did ye? You'd expect him to feel kinder grateful, maybe. But them 'Rapahoes don't care shucks fur you. If yo're kind to them, they think yo're scairt. No, sir. The only way to handle them Injuns is the way you'd treat any other crawlin' varmint. Hit the son-of-a-bitch over the head with the bar'l of yore gun the fust time you see him, and after that he'll respect you.'

They went on, watchful, past the Coon Creeks, Paw

nee Fork, Walnut Creek, past the spot where Kit had helped with the amputation two years before, plodding steadily back to Missouri, back to Franklin, back to the saddler's shop! Kit grew more discouraged every step of the way. "Pears like I'll never git to the mountains at all at this rate,' he thought. But again he met a caravan bound for Santa Fe, and once more he sacrificed his wages to throw in with the westbound train.

Back they plodded, across the Arkansas, across the Cimarron Desert, and camped on the Cimarron at last to rest for a day. There Kit saw the grave of two young traders, McNees and Monroe, a grave dug up by the wolves, so that the blankets in which the bodies were wrapped could be seen through the protective covering of poles and rocks. It was said that the two had ridden ahead of their party, as horsemen were constantly tempted to do. They had halted on McNees Creek, as it was called now, and had carelessly gone to sleep there. When the slow-paced oxen came up with them, their comrades found one dead, the other dying, shot with their own guns. . . . Not far off on the prairie lay the bones of six Indians, victims of the vengeance of the trading party, who had shot the first redskins they met. What trouble might follow no one could guess. But the men of the caravan talked seriously among themselves. Whatever the rights of the matter, one thing was certain. There was no faith to be put in Indians. Eternal vigilance was the price of safety.

Back in Santa Fe, Kit was again without work. One Tramell, a trader, hired Kit as interpreter, and took him along to Chihuahua with him. It was a long journey of more than five hundred miles, and uneventful.

Chihuahua was the biggest city Kit had ever seen, and the southernmost point of all his many wanderings. There he was paid off, and, finding Robert McKnight working the Gila River copper mines, hired as teamster and packer. In the spring the old ambition revived, and, taking his pay, Kit set out for Taos again. It was a long way back this time, and he arrived late in the summer of 1829, tired, broke, but home again.

As usual, there was no work for Kit Carson in Taos. He hung around for a few days, hoping against hope. Then he realized, suddenly, that his one chance lay in making tracks for Santa Fe. He must catch the last of the caravans bound for the States. And this time he knew he would have to go all the way: all the way to Franklin. For there could be no turning back with a westbound train. No trains came westward so late in the season. And he must go at once. The matter was pressing. He would have to raise some cash. And he had only one thing of value — his father's old flintlock. Maybe Ewing Young would give him a few dollars for it.

Under the *portal* before Ewing's house he found a group of mountain men. They were listening to an angry speaker, wagging their shaggy heads, shifting uneasily on their moccasined feet, spitting tobacco juice, grumbling and swearing, deep-throated. It was them cussed Apaches. They had licked Ewing's brigade on Rio Gila, druv 'em back to Taos. And the worst of it was, as everybody knowed, the darned Spaniards wouldn't do nothin' about it. Under Mexican law, the brigade had no business to be trapping on the Gila. No licenses were issued to *gringos* to trap in Mexican territory.

Kit pushed through the group, through the familiar doorway, into the warm gloom of the American House, up to the bar where Young stood talking busily. Kit was in a hurry. He laid the long rifle on the bar, looked Young in the eye, and spoke. 'I reckon I've got to push on to Santy Fee, Mr. Young. I need a leetle money. Kin ye use this hyar rifle?'

Young looked at the sturdy youngster with the cold blue eyes and the steady hands — at the rusty rifle with its hickory wiping-stick and the ancient notches on the smooth brown stock. He thought he knew a man when he saw one, and he needed all he could find just then. Maybeso Kit would help him wipe out them Apaches.

'I sure kin, Kit. And I kin use you, too. Never you mind Santy Fee. Come along with me, an' I'll larn ye to raise hair and set trap for beaver. How about it? You don't want to drive team no more, do ye?'

His chance had come! But Kit concealed his excitement in an even slower drawl than usual. 'Wal, I aint stuck on the job, if it comes to that. I'll go with ye. . . . But I'll need a new skinnin'-knife,' he added.

'Sure,' came the answer.

From the shelf behind him, Young took a bright new blade and laid it on the bar under Kit's nose. Kit took up the knife by its bone hilt, fingered the incised trade name *Green River* on the steel just below. He knew the approved make, though he could not read the letters. That bright blade shone with all the realization of his dreams come true. He was a mountain man at last.

The others stood and watched him, somehow aware that the moment was important. Kit glanced at them, then passed the ball of his thumb over the sharp edge of the knife. He looked inquiringly at Ewing Young. It would be war — up to the hilt — up to *Green River*. Wagh!

'Whar's the grindstone, Cap'n?' he asked.

II
MOUNTAIN MAN
CHAPTER IV
THE FIRST TRAP

SCALPING, the best authorities agree, is a European refinement upon the native American sport of head-hunting. The utter savagery of the Indians profoundly shocked the British officers. It was annoying to receive a basket of French or Colonial heads in a state of imperfect preservation as often as the warriors visited the frontier posts. It seemed hardly good form, this collecting the heads of one's enemies — even though it was necessary to encourage the Indians to kill them. Things were rather at an *impasse* until some one hit upon the happy expedient of peeling the trophy and bringing in the hair instead of the head. Such a scalp could be carried with ease in any quantity; it could not be duplicated from the same victim; best of all, it would keep indefinitely. The good taste of the officers of the Crown was vindicated: thereafter, only hair was acceptable, only hair was paid for.

The settlers found the new custom much to their liking. They maintained it. Not a few commonwealths still have on their statute books laws providing for bounty on Indian hair. A very little reading in the records of the early settlers of the States will demonstrate how universal the scalping of Indians by white men was at a certain stage of pioneer civilization. The settlers fought the devil with fire. Probably as many Indians were scalped by white men as whites by Indians, if the facts were known. In the long run the white men were better armed and more numerous than the Indians, and war was not — to them — a pastime.

War was a pastime to the Indian, and he was rarely very much in earnest about it unless inspired by bitter feelings of revenge for recent injuries. Left to themselves, the redskins hardly suffered more damage in their usual affrays than a football team does in a season. The object of war was not to inflict death and wounds upon the enemy, but rather to display courage, win renown.

The custom of 'counting *coup*' illustrates this. The word *coup* was borrowed from the French *coureurs de bois*, but came to mean more than a mere blow, stroke, attack. By that word or its equivalent, the Indian of the Plains meant striking the enemy with the hand or with something held in the hand either while the enemy was alive and active or just after he fell. To count *coup* was to strike such a blow, or to describe, in public, just how such a blow had been struck. Such rehearsals were a regular feature of every tribal gathering, for every man's prestige was simply the total of his known *coups*. He therefore lost no opportunity to recite them, and never missed a chance to demolish any false claims to such honors on the part of his comrades. To count *coup* was to win honor.

Killing an enemy with a bow or gun at long range was no honor at all, though it might be useful to the tribe. The earliest battles of the Indians had been fought hand to hand, simply because they had no long-range weapons. Afterward, when missiles came into use, it was still felt that they were a little dishonorable, a little cowardly. It was still believed that a brave man would grapple with his foe, would come to close quarters. And so, although the shooting of an enemy was creditable, useful, and all that, it was not rated as a *coup* proper, unless the slayer rushed up and touched the body as soon as it fell.

The scalping of an enemy, also, was not an honor, unless it could be counted as a *coup* — that is, unless the scalper was one of the first four warriors to touch the body. Four men could count *coup* on one enemy in the same fight, but not more than four. Each one, as he ran up, called out, 'First,' 'Second,' 'Third,' and 'Fourth,' as his turn came. And often there were very spirited races to see who should first count *coup* upon a fallen foe. Often brave men carried only a short stick into battle, desiring to show how brave they were — brave enough to fight without weapons, like officers going over the top with only a walking-stick. Such sticks, decorated with feathers, were known as *coup-sticks*.

Scalps, to the Indian, were merely trophies, something to take home to show the folks, something to talk about and dance over where the women could see. Scalps were not honors, nor were they long preserved, as a rule.

The coming of the white men changed all this. They cared less for honor than for victory, and they had no sentimental objection to killing an enemy at a distance. In fact, they preferred it. War was no game of tag to them. Moreover, they fought against odds, outnumbered. They fought to save their skins. Rarely could they escape by flight. The red man, on the contrary, habitually fought and ran away to fight again some other day. For when he ran away, it was next to impossible to catch him. War was a pastime to the Indian.

For all that, Indian wars were fearfully sanguinary affairs. On both sides the fighters were hunters, men who lived by the daily use of skinning-knife and rifle, trained shedders of blood. Our law provides that a butcher may not sit on a jury in a trial for murder, because he takes bloodshed too casually. But in the Old West every man was a professional butcher. Blood and wounds were matters-of-course to him. He killed something every day.

In these savage encounters, the white men, outnumbered as they were, had the advantage of their civilized habit of coöperation, their power of organization. They could depend upon each other. The Indian, on the other hand, fought as an individual, when the spirit moved him, and believed in dreams, omens, luck, 'medicine.' His courage waxed and waned with every mood, with every change of the moon. When he thought his luck was bad, he simply faced about and went home to his lodge-fire. Only a fool would throw away his life: no warrior could be expected to commit suicide. And so an Indian charge inevitably split on a steady group of mountain men. The leader of a war party was disgraced if he lost even one of his warriors. Hence an Indian made an uncertain ally. Nevertheless, nothing can be more obvious than the terror he inspired. It is written between the lines of every contemporary record of Indian warfare. When luck was with him, he fought gayly; when he was cornered, he fought like a wildcat in a trap, giving no quarter, expecting none.

The trappers believed that the Indian could not be trusted. But they also believed that he could be intimidated — if enough blood were shed. To the Indian a *coup* was an honor: to the trappers a *coup* meant a good Injun — a dead Injun — an Injun good and dead. Their practical point of view is well shown by the means they used to record their *coups*. Brass tacks hammered into the stock of the rifle marked the tally of the mountain man's victims. Brass tacks.

Ewing Young let every one in Taos suppose he was headin' for the States. He led his forty mountain men on the trail to the Raton Pass and the Arkansas River, then the boundary between the United States and Mexico. But after fifty miles or so, as soon as he had left the Mexican officials safely behind, he turned his face to the southwest, back to the country of those Apaches who had bested his last brigade. 'We'll see who has a right to this country,' said the Captain.

The forty trappers rode with him, a motley cavalcade, the French Canadians gay and garrulous, the Americans grim and taciturn as usual, though ready to burst out into furious spasms of rough mirth or profanity. Their ponies moved steadily along at rack or Indian fox-trot, and the pack-mules scuffled after through the dust, to the movement of flopping possible sacks and the clink of traps. Most of the men wore blue shirts, long, full-skirted buckskin hunting-coats, leather breeches and leggins, moccasins, slouch hats, or bandannas turbaned round their heads. All carried rifles, sheath-knives, bullet-pouches and powder-horns, whetstones, cases for pipe and tobacco. A few, more completely Indian than the rest, wore gee-strings and leggins that left their buttocks bare.

Some had pistols in their belts, some hatchets. They pushed on day after day, passing through the country of the Navajo, passing the pueblo of Zuñi, where Kit looked in vain for the 'white Injuns' fabled to inhabit there. At last they reached the head waters of Salt River, here swift and fresh and wooded — quite unlike the brackish stream it would be below the salt beds downstream.

Soon after they made camp, Jim Higgins saw Injuns on the skyline, and Ewing Young made preparations to receive his guests. He knew he would have to entertain them in his camp. The Apaches were in great force, and full of the pride of their late victory. Young had no desire to stand a siege. And the Indians, as usual, wished to have a look round before they attacked the party. They approached, and Captain Young beckoned them on.

At the first alarm, Kit and three fourths of the other men were ordered to conceal themselves. Before the Indians had time to estimate the number of whites, the ambush was laid. Young and the boldest of his men stood in the midst of their packs to receive the redskins, who poured down from the hills on every side, expecting to reconnoiter at their leisure, to intimidate the trappers, confiscate their stock, perhaps kill them. Down they came — at first a few — then more, until half a hundred stood in the middle of the encampment. Then the Captain gave the signal, called out, fired.

One moment Kit had been staring out of his hiding-place under his packs at the huddle of insolent brown men with their sinewy, naked bodies, their long white breech-clouts, their bobbed black hair and cloth headbands, their queer yellow moccasins reaching to the knees and having the turned-up toes. The next moment he was drawing a bead on the breast of the nearest brave, pulling the trigger, feeling the powerful kick of the old flintlock, the smart of smoke in his eyes, the roar of forty rifles in his ears. He heard the cries, the oaths, the glee of the mountain men as the Indians fell, as the survivors ran. As the smoke cleared, he saw on the ground among the packs the strangely distorted brown bodies, the blood, the bows and lances dropped. . . . Up on the hills the remnant stood yelling and shaking their weapons. Up the slopes ran a scurry of naked, dodging Indians whom the mountain men, coolly loading and firing, were doing their best to bring down.

Afterward Kit inspected the Apache he had killed. A fair shot — straight through the nipple at which he had aimed — straight through the heart within. Following the example of his companions, he got out his knife, and cutting a circle round the dead man's crown, yanked off his topknot slick as a whistle. It was his first *coup*, and rather proudly he hammered brass tack number one into the old brown stock above his father's ancient notches. And all around him was the obscene merriment of men who had just killed, men who had only a moment before been scairt of losin' their own hair. He himself was quite contained, being new to danger. He had aimed and fired with all the deadly singleness of purpose that belongs to nineteen years. But now he too felt a little queer, a little shaky, ready to pack up, take a nip of Taos Lightnin', and push on.

His reward came when he overheard the Captain

say, 'Kit thar's a likely young un. He'll make a mountain man yit.'

Young and his men trapped down the Salt River and up its tributaries, taking great quantities of furs. Sometimes they found dams which could be broken down and the beaver killed with clubs and hatchets; sometimes they had to use all their finest skill in wading, placing their traps, using the luring castor to outwit the cunning creatures. So much fur was taken that the Captain decided to send back twenty-two of his men with the catch, and to lead the remaining eighteen into California and up the Sacramento. It was always the farther pasture that lured the trapper, and the Indians had become very troublesome since the late massacre. Nearly every night they crept into camp, stole traps and packs, cut the throats of horses and mules, made themselves generally a nuisance. When the news of the intended division of the men got about, Kit Carson's heart sank. He was the youngest. More'n likely the Cap'n would send him home with the packs.

But when the names were called, Kit found himself among the chosen eighteen. He was making good. Already he was rated as belonging to the better half of the men. It was his first real recognition. For every available white man in Taos had been asked to accompany the expedition.... But now! Now he was Mountain Man! Wagh!

The eighteen trappers pushed on across the desert. They had hunted and jerked the meat, so as to have provision for the dry scrape. They had made tanks of the hides of the deer they killed and filled them with water for animals and men. But with all their precautions, they suffered very much. For four days they had to drink from the tanks, had to post sentries to keep their water from being used up by the thirsty men. The fourth day the mules, strung out along the road for miles, smelled the water ahead, began to trot, led the men to the first pools. Four days later, after another dry desert march, they reached the brim of the Grand Canyon. Kit and his friends were among the first white men who looked down into that sublime chasm, and it stirred them profoundly. For down at the bottom, a mile below, they saw the thin ribbon of the river. Water! Running water!

The men were starving for fresh meat, too. They found some Mohaves, and from them purchased an old mare, some corn and beans. Kit's own story tells how the mare was 'heavy with foal.' It 'was killed and eaten by the party with great gusto; even the foal was devoured.'[1] Peters, the official biographer, omits all mention of the foal; he could not share the Indian's enthusiasm for that dish. But when hunger pinched, nothing came amiss to the trappers. Their custom followed that laconic proverb of the Old West — 'Meat's meat.'

Leaving the Grand Canyon, the party marched up the Mohave River, a stream which has the odd desert custom of flowing upside-down, with the water under the sandy bed. Leaving this stream, four days' ride brought them to California and San Gabriel Mission. There they found a thousand Indians, fifteen soldiers, one priest — a staff which well indicates the comparative importance of religion in those outposts of the Holy Faith. After a short rest, they passed San

[1] Grant, *Kit Carson's Own Story of His Life*, p. 14.

Fernando Mission, trapped along the Sacramento and the San Joaquin Rivers, and at last ran into a party of Hudson's Bay Company trappers under Peter Skene Ogden. As game was plenty and beaver scarce, Young and Ogden decided not to start hostilities, and for some time the two parties remained in the same neighborhood on friendly terms. Then Ogden started for the Columbia, and Young's men spent the time hunting on the Sacramento.

The Mission of San José had a number of Indian malcontents among its converts. These ran away and took refuge among hostile Indians. The *padres* asked Young to send his men after the runaways. Young was willing to oblige the Mission in whose country he was hunting without invitation, and sent Kit with eleven men to round up the fugitives. Some of the Spaniards went along. They had been defeated by the Indians in one skirmish.

Kit reached the hostile village, made his demands, and, finding these rejected, attacked. The Indians lost heavily, retreated, and Kit burned their town. When he repeated his demand that the deserters be given up, the Indians brought them in. It was their first experience of mountain men in action, and they realized that these were no Spanish soldiers. Kit returned triumphant, without the loss of a man. It was his first independent command, his first military victory. His heart was big that day. Young and his men became favorites of the *padres*, and were soon enabled to sell some of their furs to the captain of a trading schooner then at the Mission.

One night not long after Indians stole sixty horses from the camp. Kit was sent with a dozen men to recapture the herd and punish the thieves. After a ride of a hundred miles, he found the Indians feasting upon the flesh of some of the stolen animals. Kit led a charge, killed eight warriors, and rode home with most of the stock. It began to look as though Captain Young could not get along without his right-hand man, Kit Carson. Whatever he was sent to do was done most thoroughly. Kit was eager to make good. He had not starved three years for nothing. He wanted no more cooking, no more teaming, no more envious idleness in Taos. The boy's ambitions were being realized.

Now the trappers were flush with Spanish silver, and in spite of Young's objections insisted upon stopping in the pueblo of Los Angeles for one glorious spree before crossing the desert again. Discipline was hard to enforce among the trapping fraternity. They came of the Kentucky and Ohio pioneers, men who moved on as soon as neighbors made clearings within the sound of their axes. They were not soldiers, not wage-earners, but lone adventurers bound together only for better protection and because they owed money to their *bourgeois*, their boss. Lone nomads, their only relaxation was the fandango, a deck of euchre, whiskey straight. Young compromised — and went to Los Angeles.

They clattered into town, eyeing the women, prepared for a wild night. But Los Angeles was not a Mission, and the authorities at once demanded Young's passports. Passports! It made those shaggy wanderers laugh. Their rifles were all the passports they required. Bring on the *señoritas*, the Spanish brandy, the music. Damn the lousy *pelados*. 'We're on our spree. Hyar's the beaver!' And they rattled the silver under the noses of the darned Spaniards. The officials laughed too, and began to sell liquor.

Reckless, youthful, tough on principle, lonesome for some gayety, for some human society, the mountain men accepted the fire-water with open throats. They swallowed it by the gourdful, and very soon were as drunk as only trappers could be. Young was in despair, for he saw what would happen when all had become too drunk to fight. They would all wake up in the calaboose to find their furs confiscated — all those packs of priceless furs. Once in Santa Fe the Governor had mulcted him of peltries to the value of twenty thousand dollars. He had no intention of enduring such robbery twice, if he could help it. He looked round for help, found Kit still sober, gave his orders.

Kit took three men — the only ones who could be prevailed upon to go — all the loose stock, the packs, and camp equipment, then started for Taos. Young told him to push on as fast as he could. The Captain stayed by his drunken men, trying to get them on their feet, into their saddles, trying to get them headed for Taos, away from the dangerous town of the Angels. Young had his hands full, and as the trappers staggered out of the *sala*, clambered into their saddles, swearing and singing, maudlin under the stars, the smiling Spaniards followed along, offering more liquor, ever more liquor, waiting until their chance should come. And the mountain men drank all that was offered them.

Their own cussedness saved them. For one of the men, poor Jim Higgins, blind drunk, walked deliberately up to his friend, Big Jim Lawrence, and shot him in cold blood. That frightened the Spaniards, sure enough. *Dios!* If the mountain men murdered each other so casually, what might not happen to a native, once the fur began to fly? In a few minutes not a Spaniard was to be seen, and Captain Young led his men safely away and overtook Kit next day. Big Jim Lawrence recovered, owing to the extremely drunken aim of his friend, Jim Higgins.

On the way back the party heard the thunder of hooves one evening, saw a great herd of horses running, with mounted Indians behind and on the flanks. Stolen stock! With their usual eager desire to take part in everything that happened, they charged the Indians. The thieves ran, and the herd fell to the trappers. Casually, they took over the animals, culled out the best, let the rest run. Horseflesh was a temporary possession at best in the Old West. Peters naïvely remarks that 'to return the animals to their owners was an impossibility.' The fact is, such a notion can never have entered the head of a mountain man. Return the horses to a Spaniard, indeed! Hell's full of Spaniards!

They trapped all the way back — the Colorado, the Gila, the San Pedro. One or two skirmishes with Indians occurred, and once Kit had to help bluff a band which threatened to rub out the party in camp. His knowledge of Spanish was useful that day.

At the copper mines on the Gila, Young's trappers met up with Robert McKnight, Kit's old employer. With his connivance they concealed their heavy packs in his old diggings, and Young sent his men on to Taos. Kit and he rode to Santa Fe, where Young got a license to trade with the Indians on the Gila. Returning to the mines, he brought in his packs, sold them openly in the plaza at Santa Fe for the market price — two thousand pounds of beaver! Every one remarked upon what a handsome trade he had made in so short a time! Young packed his silver on mules and set out for Taos. With the furs already disposed of at Taos and in California, the proceeds of the expedition amounted to some twenty-four thousand dollars. A tidy sum in 1830.

And now we see Kit Carson, Mountain Man, riding into Taos on a fine red California mule and silver-mounted saddle. He wears a brand-new hunting-coat of golden-brown buckskin, fringed and beaded, with patterns of quill-work on the sleeves and diapers of black velvet let into the full skirts. The coat has a fur collar, brass buttons. He has on new Ute moccasins. From among the fringes of his leggins dangle tufts of dark hair too fine to be horsehair. He sports a shiny new rifle with three brass tacks in the stock. On his head is his first genuine beaver hat. The big spurs clink, the mule wags its long ears to the tune of the jingling Spanish bit, its hind-quarters heavy with the silver in Kit's saddle-bags. The boyish freckles have disappeared under a coat of heavy tan. And the man's blue eyes are fixed on the trail ahead.

Ahead, on its hill, the white walls of Fernandez de Taos shine in the sunset. To-night there will be a fandango in somebody's big *sala*. To-night the trappers will caper and howl. To-night Taos Lightnin' will go down lusty throats by the gourdful while *bandolin*, guitar, and tom-tom make the blood flow faster. To-night Kit Carson, Mountain Man, with money in his pocket and a heart as big as Thunder Mountain, will swing the smiling *señoritas* in their short skirts and thin white blouses, swing them under the noses of their Mexican lovers — spiritless *pelados*, greasers! For the *señoritas* will be waiting, fat and sassy and ready for huggin'. To-night Kit Carson will be on his spree!

Whoopee!

CHAPTER V
KIT CARSON, TRAPPER

By the end of summer, Kit was broke again. Fandango and *aguardiente*, euchre and *monte*, treating his comrades and buying fofurraw for the Taos girls had drained his pockets as a break in the dam drains a beaver town. Gone was the red mule, gone the silver-mounted saddle, and his possible sack was limp with its emptiness. Kit was hard-up, a hanger-on at the American House, one of forty impoverished mountain men in Taos. Ewing Young could afford to be generous: he had all the furs and all the money paid for them as well. Occasionally some of the men went into the mountains hunting and brought back a little meat or a few deerskins. But that was only to escape boredom, a mere pretense. They lived on credit or on the free-handed generosity of the Old West, waiting for the fur season, waiting for autumn, for the fall hunt. There was no work in Taos, and few of those aristocrats of the beaver stream would have stooped to do it if there had been.

Autumn days came at last. Snow appeared on the mountains, and bright splashes of yellowing aspens spotted their sides. Kit and the others were restless, grease-hungry for fat cow and mountain doin's, half froze for the trail and the chance of a Pawnee topknot. Even the simple civilized fixin's of primitive Taos were too effete and fofurraw for the trappers. With what savage gusto they spat out their pronunciation of the

French word — *fanfaron* — that synonym for everything frivolous, showy, effeminate, unessential, belonging to the despised settlements. Fofurraw! Everything this side the mountains was too damned fofurraw!

When Thomas Fitzpatrick, partner in the newly formed Rocky Mountain Fur Company, rode into Taos looking for recruits, the eager mountain men enrolled in his service *en masse*. Fitzpatrick stood behind Ewing Young's bar and laboriously entered the names and debits of his men as each one pushed forward in turn to claim his equipment. The little room was jammed with the candidates. Broken Hand (as old Tom was called because of an injury due to the explosion of his gun) had the respect of every man present. They were glad to have him as their partisan.

Kit was pushed forward in his turn, and stared at his new boss. Tall and muscular, with sanguine complexion showing through his tan and thick beard, Thomas of the Broken Hand looked at Kit with piercing eyes from under the brim of his old wool hat. Those melancholy and uncompromising eyes seemed to have but a poor opinion of the stocky, sandy-haired young 'un who looked 'most too small to set a trap. But Ewing Young said a word in Tom's ear, and the new partisan took the quill between his shattered fingers and made entries in his book. The men watched in fascinated silence. To Kit, those illegible marks were significant. Now at last he was to head into the Rocky Mountains, the trappers' paradise.

Broken Hand scratched awkwardly in the small black book, reading each entry aloud as he proceeded:

KIT CARSON

Dr.
Sept. 1830. To.... 1 saddle mule.......... 30 plews
　　　　　　To.... 1 Spanish saddle........ 40 plews
　　　　　　To.... 1 capote, Hudson's Bay . 8 plews
　　　　　　To....Galena lead pigs........ 1 plew
　　　　　　To....3 feet Twist tobacco.... 1 plew
　　　　　　To....6 traps................. 24 plews
　　　　　　　　Total..................104 plews

'Next,' said the Broken Hand, and Kit pushed his way out to get his plunder together for the long ride north. It would not take long to pay off that debt, once he got among the fur. Had he not overheard Ewing Young's whisper to the partisan: 'Yon's a likely beaver kitten, Tom?' ... A plew, or, as the French had it, *plus*, was trapper slang for a prime beaver skin. But long usage had made it an arbitrary symbol of value, the coin of the mountains. Kit saddled up that afternoon and rode out of Taos with a few others, having told the partisan of their intended movement up the trail. Kit was eager to be on his way.

Next day Broken Hand got the party together and started, after considerable cussin' and rasslin' with refractory mules. They rode east from Taos through the hills, then turned north toward the Arkansas River. Their way lay along Greenhorn Creek and the Rio San Carlos, and they forded the Arkansas near the mouth of Fountain Creek, where the city of Pueblo, Colorado stands now. Following Fountain Creek northward through the barren hills, Broken Hand's party crossed the ridge, rode down into the valley of Cherry Creek, and so came at last to the Platte. Trapping its tributaries as they went, they steadily moved along, reached Sweetwater, Green River, and Jackson's Hole. Somewhere in the Rocky Mountains there should be a Carson's Hole. But there is none, for he was too constantly a wanderer to associate his name with any of those lovely sheltered valleys of the ranges, those 'holes' or winter retreats beloved by the early mountain men. After all, Taos was Kit's hole, and no park or valley of the mountains he loved so well will bear his name forever.

When winter came on and the streams froze and the beaver retired to his lodge to hibernate, the trappers went into winter quarters on Salmon River. There they built their arched framework of saplings, covered with skins, and open on one side to the blazing fire. There they constructed their scaffolds for curing meat, and rigged a rack of poles or antlers on which to hang their fresh meat and traps and clothing. There were placed the graining blocks and stretching frames used in dressing the skins and peltries. Inside the hut were stowed the saddle, the packs, the buffalo robes and blankets for the bed, possible sack, rifle, hatchet. In the bottoms the mules grazed on the good grass or, when the snow was deep, gnawed the bark from cottonwood boughs cut down by their owners. It was a satisfying life. Game was plenty, for the buffalo came in swarms from the freezing plains, and the men feasted on fat cow and prairie oysters.

Kit was well pleased with his fall hunt. He had done well, and was becoming a really expert trapper. Already he had paid his debt to Broken Hand, and had besides two full packs of skins, each weighing a hundred pounds, and containing eighty beaver plews each. He had picked up some otter, fox, and muskrat skins as well, and had a bear rug to add to his blankets at night when the fire burned low and the wolves howled from the windy hills. He was enjoying the prestige also which belonged to the man who was a genuine *hivernan* or winterer.

That winter, while the beaver tails boiled in the pot to delicious tenderness and the aroma of coffee blended with whiffs of strong tobacco from his old black clay, Kit and his comrades would hear Jim Bridger vie with the best in telling the whoppers beloved by the mountain men.

'Why, boys,' Jim would begin, solemnly, rolling his eyes and spitting deliberately into the embers, 'you know I fust come out with Ashley. Wagh! When I come west, yon hill was only a hole in the ground. Off-hand, or with a rest, I make 'em come. Thar aint a varmint in the mountings thet I hain't et. I knows sign when I sees it. . . . But oncet I seen something thet made me feel almighty queer, I tell ye. I war trappin' in the Black Hills. It war cold that day, fit to freeze yore breath in chunks. And up on the peaks where the snow had been a-meltin', the water purty nigh froze my feet off. But as I come a-wadin' down hill, purty soon I noticed that the water war gittin' warm, and swift as all get out. I looked ahead and seen steam. Boys, I tell ye, I jumped out o' that crick quicker'n scat — and jest in time. Why, damn my hind-sights if that water hadn't run down hill so fast it was a-bilin' at the bottom!' . . . And 'old Gabe,' as his comrades called him, would look round, solemn as a totem pole, wagging his beard at the crowd.

Then some other would enter the arena, and Kit would hear of the giants who lived on an island in the

Great Salt Lake, giants who built immense log houses and ate corn from cobs a yard long. What folks thought was forest fires was only the smoke from their pipes. . . . Kit would hear about the crystal mountain so clear that nobody could see it, so clear that its location and size could only be guessed at from the stacks of bones of animals and birds which had broken their necks by running into it; about the echo in the Bighorns which took eight hours to return, so that a man on going to bed had only to yell 'Git up!' and next morning promptly the echo would rouse him with a loud 'Git up!'

And some one would narrate the one about the time Bob Hatcher rode under the Spanish Peaks straight into hell and played poker with the old Black Bear himself; about how Broken Hand killed three buffler with one bullet; about the 'putrified' forest where putrified birds sang putrified songs; about the devils who jump up and down in the boiling springs at the foot of Pike's Peak; about the time Jim, or Tom, or Kit shot his mule in mistake for an Injun; about the time old Cotton was cornered by the Blackfoot, with no chance of escape. Old Cotton would dwell upon the hopeless situation. Then he would pause, waiting for his listener to bite. 'Well, what happened?' would come the question. Old Cotton would spit deliberately, roll his cud villainously, and would answer, before exploding in a loud guffaw, 'Why, dang ye, then the Injun killed me!'

The point of the whopper was always that the narrator himself was its hero. No mythical Paul Bunyan could ever have been created among the free and independent trappers. No wage-slaves they, to make a hero of their employer. On the contrary, the *bourgeois*, the boss — with few exceptions — was regarded as quite contemptible. With true Indian brag, each trapper made himself the center of every yarn, however wildly incredible. Yet the joke lay, for these utter realists, in the salty contrast of this crazy individualism with the sane common sense of their breed. Their whoppers always appeared perfectly reasonable, perfectly logical, except in one point — a point more logical than all the rest, yet utterly absurd. Their endless lies, their yarns, were the literature of their contempt for the order of the settlements, a justification of their escape to the wild life of the mountains and the plains, where the vastness and haphazard chances of life made logic look a little ridiculous, a little fofurraw.

'Sure, it froze my hair till it was stiff as wire. Sure. I put my hand up to scratch my head, and damned if it didn't come away as full of spines as if I'd stuck it into a porkypine's back. Tuk me all night to git them hairs outen my fist.' And the narrator would scan the back of his hairy paw anxiously. 'Look hyar. Dang me, if thar aint some of them hairs still thar! Don't they look the same color as the hair on my head, now?' And the listener would have to guard himself from falling into the trap prepared, so serious was the air of the spinner of yarns.

And then the favorite whopper about the mosquitoes, and how they drove the trapper into his tent. Even there the pests pursued him, pushing their long bills six or eight inches through the canvas. 'And so I thinks to myself, "I'll fix ye, damn ye." And I picked up the axe and went for 'em. Clinched all their bills on the inside of the tent, jest as if they war nails, see? Then I says, "Now, durn ye, buzz yore belly full; I'm goin' to sleep." So I turned in, but purty soon I woke up agin to find the devils bitin' me. And away up yonder against the stars I could see my tent a-sailin' away to the mountings. Durn my skin if them mosquitoes hadn't flew plumb away with tent, and poles, and all!'

That winter Kit made friends. Friendships in the mountains were real friendships. The trappers were few in number and much together. Under the conditions of their life, no one could pretend to any qualities or virtues he did not have. The shirker, the coward, the braggart, the stingy — not one could impose on his comrades as being other than he was. The day's march, the night attack, the needy friend — all these afforded constant tests for the character of the trapper. A man's worth was soon as definitely known as his weight or his inches, because as readily ascertainable. They led a hard, dangerous life, and never was there any doubt whether a man had met those conditions or not.

Once he passed the tests, he suddenly found himself mysteriously a member of a fraternity, already a beloved comrade. And Kit Carson, hearing the men address him in terms which, in themselves, were savagely profane and insulting, knew that he had passed the tests, knew that they were caressing him in the only endearing terms they knew. In the settlements he would have resented those foul epithets, would have fought the man — however big — who used them. There no one would have been allowed to refer to him continually as a child, a nigger, a horse, a lousy skunk, a son of all manner of evils. But here it was the mark of friendship to be so called — by the right people.

And Kit soon learned in other ways that he had graduated from the elementary school. Blackfoot Indians attacked and killed four of the trappers while they were hunting. Blackfoot! A name of terror to the mountain men, though hardly worse than Arapaho, Kiowa, Comanche, Pawnee. Plains Indians — proud fighters.

No Mission slaves these, to run away and let him burn their town. No miserable Diggers, no skulking Apaches crawling into camp at night to steal a trap. Blackfoot! ... Men talked of how them cussed Injuns had rubbed out others of their acquaintance, of how many a trapper had disappeared into the mountains in their country and had never been heard of again. Kit fingered the three shining brass tacks in the stock of his rifle and wondered how soon he would be called upon to fight. Could he add to those three *coups*, or must he go under, rubbed out by the redskins? Plains Injuns were upstanding fighting men — most of them a whole foot taller than he was. Why, them Blackfoot would fight all day — with white men — mountain men — right out in the open! Wagh! It warn't hardly natural. ... But anyhow, here in the mountains it was better than out on the plains. The plains! That was where you'd catch it!

In April Kit and his comrades set out on their spring hunt. Now — after the bitter winter — the beaver fur would be at its best. Now their saddle animals, scant of hair but seal-fat after their winter's gnawing on cottonwood bark, were full of the old ginger and rarin' for the trail. The men loosened their white blanket coats, threw back the attached capotes, and rode away with heavy packs, trapping the tributaries of Bear River, Green River, in small parties. Kit ran into some of Sinclair's men, learned that Captain Gaunt was in the New Park, decided to take his wages and join him. Kit had seen Gaunt in Taos.

Gaunt gladly welcomed Kit and his comrades. Together they trapped the New Park, the lonely Laramie Plains, the South Platte, and the headwaters of the Arkansas. On the Arkansas they went into winter quarters while Gaunt took their furs to Taos and brought back supplies.

Buffalo were plenty that winter, and Kit's party killed a great many, running on snowshoes alongside the shaggy animals wallowing through the deep drifts, shooting, lancing, stabbing the helpless beasts as easily as so many cows in a barnyard. The robes were in prime condition, and the meat soon disappeared, what with all those trappers and Indians to feed.

All about Gaunt's camp were villages of Arapahoes. Their lodges were filled with fine buffalo robes, robes which Gaunt was eager to get. He tried to trade with the Indians, but to no purpose. All he had to trade was whiskey, and after one gasping gulp, the chiefs utterly refused to drink the fiery stuff. 'White man's water no good. Taste bad. White man no catchem robe.' And that was final.

Gaunt was stumped for a while. But the thought of losing those robes kept him awake nights. He had a little sugar on hand — a treat for which the Arapahoes were willing to pay heavily — one dressed robe for a half-pint. But the sugar would not go far, and he had nothing else to trade.

One day a happy thought struck him. He was sitting in the lodge with several chiefs, smoking the long pipe with them. He laid the pipe aside, beckoned the chiefs to watch him. He poured out a cupful of whiskey. Then, showing them a handful of sugar, he slowly poured it into the dram, and stirred it with the blade of his skinning-knife. After they had had time to think it over, he offered the cup of sweetened liquor to the principal Indian. Kit sat by, curious to see what would happen.

The chief looked doubtfully at the fiery liquid, poked his finger into it to feel the sugar on the bottom, smiled, and drank. It was fiery, but it was sweet. The chief grinned approbation, passed the cup for more. Gaunt gave every one a sample, then asked them to announce that he was in trade for buffalo robes. They strode away, warmed by the draught, and spread the news. Soon after the Indians came swarming, bringing their mules — or their women — loaded with shaggy robes. Kit and his comrades were kept busy beating and packing the robes for Gaunt. The Arapahoes, soon too drunk to know what they were drinking, were given alcohol without sugar, alcohol diluted with water, poisoned with tobacco and pepper, served from cups half filled with melted tallow to increase the profits of the trader. The Arapahoes became a nation of drunkards.

Kit Carson was disgusted with such goings-on. He knew and approved good whiskey, and he had no great love for the Indians. But Kit was an honest man, and Gaunt's shameless cheating shook his confidence in his partisan. Kit hired out to Bent and St. Vrain, whose big adobe fort was still building down-river, a few miles above the mouth of the River of Lost Souls, the Purgatoire. There he was put in charge of a gang of loggers, and went into camp on Wild Horse Creek, half a dozen miles below the fort.

Here Kit had his first real battle with Plains Indians, the fight which won him the Indian name by which he was ever after known to the red men.

The story has come down to us in several versions: in the rough notes Kit dictated to Peters but could not read; in Peters's own vague and grandiloquent account; in Cheyenne tradition.

Any one who is familiar with old-time Indians will prefer their version of a fight to that of any other eye-witnesses. For the Indian was not only a better and less imaginative observer than the white man. He was also a more interested observer, because war was his greatest and most absorbing sport. More than that, his rating in the tribe depended upon his proven *coups*, and he took good care to claim all he was entitled to, and to demolish any false claims advanced by his comrades. Therefore, in any kind of fracas he had all the keen, clear-eyed alertness of a professional sportsman. And he had the advantage of steady nerves; he was less likely to get excited than most men. He saw just what happened. And as long as he lived, whenever he counted his *coups* in public, he had to rehearse just what he had seen. For these reasons the Cheyenne tradition of this fight is preferable to any of the others. I have it from George Bent, and it has been printed by Grinnell.

One dark night a war party of fifty Crows passed Kit's camp on Wild Horse Creek. As they saw he had only a dozen men and no horse-guard, they quietly rounded up his stock and made off homeward, glad of a chance to ride, since — as was usual in winter — they were all afoot. When morning came, Kit found himself without a single head. But Kit never dreamed of letting them get away with all Bent's stock. He and his dozen men pushed hard along the trail of the thieves, which led off north across the prairie. With them rode two Cheyennes, Black Whiteman and Little Turtle, who had been visiting Kit the night before and had kept their ponies tied up.

'Twilight was falling when Carson's party, the two mounted Cheyennes still out ahead, following the trail in the snow across the prairie, saw a shower of sparks rising from a thicket some distance in front of them. The party halted and held another consultation. Black Whiteman and Little Turtle then rode off alone while Carson drew up his men in a long line, each man several paces from his neighbors on either side.

'As they advanced across the snow a dog barked in the thicket, and a moment later a little ball of white steam shot up from among the willows. The Crows had put out their fire with snow. The Americans quickened their pace and had almost reached the edge of the thicket when without warning sixty Crow warriors broke out of the willows and charged them. So fierce and sudden was the attack that Carson and his men were borne back and almost surrounded; then they threw up their rifles and gave the Indians a volley.

'Carson used to tell how surprised the Crows were when they charged in on his little party and were met by a stunning volley. Back into the thicket went the Crows and in after them went Carson and his men. The Indians evidently intended to mount and either run away or continue the fight on horseback, but when they reached their camp in the middle of the thicket they found that the horses and mules they had left there had disappeared. Right at their heels came Carson's men; so without halting the Indians rushed through the thicket and out at the far side, making off across the prairie as fast as they could go. The whites, worn out after their long march through the snow and content with the result of the fight, did not attempt to follow farther.

'When Carson had started to advance toward the willows, Black Whiteman and Little Turtle had ridden off to one side, making toward one end of the thicket; then just as the Crows charged out of the bushes the two Cheyennes rode in, stampeded the horses and mules and ran them down the creek. . . . In the morning Black Whiteman and Little Turtle returned to the thicket, and there found, counted *coup* on, and scalped two dead Crows.

'The Cheyenne have always expressed surprise that in this fight Carson and his men, all well armed and excellent shots, should have killed only two Crows . . . not one of the whites was killed or received a serious wound.'[1]

This skirmish is interesting for more reasons than one. From it Kit Carson learned some valuable lessons which he afterward put to good use. That night as he and his tired men lay on the snow, he thought things over and put away certain facts for future action. He should have known better than to let his stock run loose at night, even though it was past all

[1] George Bird Grinnell: *Bent's Old Fort and Its Builders*, Kansas Historical Society Collections, vol. xv, p. 36.

likelihood that the Crows would be a thousand miles from home in winter. He should have been ready for action at dawn, instead of having to wait to run bullets. He should have rushed the Crows, instead of letting them rush him and almost sweep his men away. His mistakes would not be repeated.

On the other hand, he had learned something about Plains Indians: how willingly they fought, whether as foes or allies. The two Cheyennes had no reason to risk their skins for Kit, but they had gladly gone along, made out the trail, captured the herd. The Crows might have stayed in the thicket safely until dark, and then have ridden away; instead they had charged into the open in the face of thirteen rifles. These were foes worth fighting, allies worth having.

One other lesson he learned, which he often put into practice. The Indians gave way when they began to lose men — that was natural enough.... But they ran like sin when they were startled by the loss of their horses. Surprise them. Startle them. Jump them — and the victory follows. Why, the Indians were just like wolves: run, and they follow; follow, and they run. That was a fact worth knowing.

Kit Carson went to sleep that night tired and hungry, but with a new confidence. He had killed his first Plains Indian. That fourth brass tack in the stock of his rifle — it counted for more than the other three.

When Yellow Wolf, that wise old chief, brought his band of Cheyennes in to trade at Bent's fort that moon, he talked over the details of the fight with Kit and William Bent, as they sat smoking together in the council room of the half-built fort. With animated gestures and broad grin he taunted the stocky, sandy-haired little white man with his failure to kill more than two Crows. And Kit, knowing what Indian decorum demanded, sat unperturbed and smiled in turn until the chief had had his joke.

Then Yellow Wolf, after a piercing glance at Kit, rose dramatically and gathered his buffalo robe about his hips. He held his chief's pipe along his left arm, and gestured impressively with his right. The white man was young, he was small, his thin hair scarcely reached the shoulders of his white blanket coat; but he was brave, too, and the Cheyennes respected bravery above all things.

The hissing, choking Cheyenne syllables began, the arm and hand swung more vigorously. 'My son, I give you a new name. You have won it. From where the sun now stands your name is *Vih'hiu-nis*, Little Chief.'

That name has stuck to Kit among the Indians to this day.

CHAPTER VI
FREE TRAPPER

IN the spring, as soon as the thaws set in, Kit and the rest of Gaunt's men went to trapping, working north toward the Laramie River. Having good store of furs and robes, they decided to *cache* them on the Arkansas until their return on the trail to Taos.

Choosing a level bench of firm soil near the river, they went to work. First the sod was carefully cut out and laid on buffalo robes spread near. Then a bottle-shaped excavation was made, narrow at the top and much bigger below the surface of the ground. All the earth taken out was carefully placed upon buffalo robes, carried to the river and thrown into the water, so that not a clod remained to show that a hole had been dug. Then the hole was lined with branches and leaves all round, and the tightly packed furs were stowed in it. A covering of poles and straw and earth tamped firm was added, and at last the sod was carefully replaced. When all was finished, no one would have supposed that the grass had been disturbed.

But the trappers were much too wary to rely upon such precautions. For a day or two they kept their horses tethered above the *cache*, so that their restless hooves would obliterate any sign of digging. After that they built a fire on the spot, so that the surface was charred and covered with ashes. Then they rode off to the north, satisfied that no prowling Indian would discover their treasure, unless he had been watching their work from behind the hills.

Now that hunting has become a form of sport, few men have any conception of the labor and hardship which the professional hunter underwent. There is no harder work than the pursuit of wild animals as a business. It is one thing to go hunting with a well-stocked larder in camp and a base of supplies within easy reach: it is quite another to depend altogether upon rifle and trap for a living. Game is here to-day and gone to-morrow, and meat will not keep. And even where game abounds a man may starve unless he is properly armed and skilled in hunting. Both trappers and Indians often went hungry. They were rarely more than three meals ahead of starvation, and quite as often five or six behind.

Kit was already one of the best of the trappers, master of a skilled trade which many a man found quite beyond his powers. Like others, Kit worked hard for the furs which he squandered so recklessly at Taos and at rendezvous. And one secret of his success was the care he used in choosing companions.

When the likely stream was reached, Kit rode out with one or two trappers, pack-mules and camp-keeper following, and looked for sign of beaver. When at length he found traces of the busy animals — a gnawed aspen tree, a cut log, shavings, a dam, a lodge, a slide, tracks in the mud — he got down and examined these more closely, making sure that the sign was recent and abundant. Then he looked for the slide on the muddy bank where the beaver played.

Near the slide he set his trap, taking good care not to leave any scent or tracks which might scare away the wily beaver. He waded upstream to the foot of the slide. Taking a trap from his trap-sack, he set the steel jaws open, placed the trap under water near the foot of the slide, and fastened the chain to a tree, or a stump on the bank, or — if that failed — to a long peeled stake in the water. Such stakes were made smooth, so that the ring at the end of the chain would move freely down them. At one end was a notch or fork. Finally he pulled the stopper from the horn bottle at his belt and anointed the end of a stick with the 'medicine' or castor it contained. This scented bait-stick he stuck up over his trap, to which he had already attached a long string ending in a marker or float. Then he carefully waded back downstream, mounted, and rode on to set another trap.

When the beaver felt himself caught in the trap, he naturally dived for his lodge. The chain slipped down the long peeled stake, and hooked itself over the fork at the end, thus preventing the beaver from regaining the surface of the water. Thus quickly drowned, it was prevented from gnawing off its foot and freeing itself, and also spared much agony. If the beaver carried the trap away, the float or marker would tell where it had gone. 'If your stick floats thataway' was trapper slang for any indication of purpose.

Having set his trap-line, Kit went into camp until that line ceased to give profitable results. Then he moved on upstream and set a new line of traps, or crossed the divide to another stream. Down that he worked to the main river up which his outfit was trapping. When in camp, he might help grain and dress the skins, though the camp-keeper generally did all such work.

Half the time alone or with few companions, often wading the icy mountain streams, and living on what meat he could kill between times, he had plenty of hard work to do, traveling miles up and downstream every day. Trapping was hard work.

Hard work. And unprofitable to the unskilled trapper, who might finish the season still in debt to his partisan. The lazy and stupid were soon eliminated, or fled back to the settlements, leaving their debts behind them. Just such a desertion took place on the Platte, where Gaunt's men were working the headwaters that spring.

One morning Captain Gaunt found four of his best horses missing and two of his men. He rode out to look for sign. It was plain. The men had ridden back south toward the Arkansas, ridden hard, leading one spare animal each. They owed money to Gaunt — though that was little enough. They had stolen his horses — and that was bad enough. But in camp that day one thought flashed into the mind of every trapper there — the furs — the *cache* on the Arkansas! The deserters would rob the *cache* — steal the labor of months — all the profits of the whole band!

The missing men were French *engagés*, *mangeurs de lard*, pork-eaters, as the trappers scornfully called them, hirelings, slaves, or peons in fact. Poor devils who hired out for less than their keep, and found the wilderness anything but a land of heroic adventure. Well, they were gone, and something had to be done about it at once.

The country swarmed with Indians — Indians to whom a stranger was an enemy. It was spring, and the season for war parties. If the thieves could get through, could the pursuers? Gaunt called for volunteers. Kit and one other offered to go, and, saddling up, set off without delay.

To Kit such rascality was incomprehensible. Why any man who had once tasted the freedom and independence of the mountains should long for the fofurraw fixings of the settlements was beyond him. If a man's stick floats thataway, he must be plumb crazy. Folks as hain't seen Injuns scalped nor don't know what a beaver trap is for — well, they just caint spit even with a mountain man, anyways you fix it. Kit hoped the thieves had been wiped out by the Injuns, and sarve 'em right!

Kit and his avenging comrade rode hard down the Platte, up Cherry Creek, over the divide and along Fountain River for the Arkansas. The trail was plain to follow, and they knew the minds of the thieves too well to need sign. It was a race for the precious furs — with the thieves eight hours in the lead. But the danger of Indians slowed Kit, and gave the deserters an even greater start, for Kit had to ride by day. That meant a halt on every hilltop and a searching scrutiny of every valley before he rode down the hill. His courage and caution marched shoulder to shoulder Kit was bold, but he was seldom foolhardy.

On the way he considered what he would do with the deserters when he caught them.

People who believe that there ought to be a law to cure everything which goes wrong can have little conception of society in an age when there is no law. The Old West, like the time of the Greek seafarings, was essentially an heroic age. It was an age of heroes, because it gave every man a chance to show the stuff that was in him. Convention, custom, was a check, but not a power. Individual desire was the spring of every action, and where all men went armed, few cared to try to impose their wills upon others. Coöperation, where it existed, was for mutual interest. The men of those days were pugnacious, passionate, and gave rein to all their impulses.

Instances could be multiplied of Indians and trappers who, like Achilles, refused to take part in the battles of their friends and rode off to look out for themselves. Gaunt's men might leave. Who cared? But the theft of the horses and the furs — *that* was unpardonable. What should Kit do about it when he caught the thieves?

Kit Carson could not imitate the modern shirker, who passes the buck to the policeman and the courts, and then wonders why criminals go free. Nor was he a gunman lacking the guts to face armed enemies, nor a sentimentalist lacking the guts to kill them. He had to decide and act at once, and he had only two guides — Shame and Blame. Shame which he might feel for an action he believed unworthy: Blame which he might feel expressed in the words or attitude of his companions. Powerful guides both — among men who cared little for property, whose enterprises depended upon mutual labor and mutual trust, men whose rating was simply the known qualities they had displayed in stress of danger. Nobody whose aim was to make money would have been a trapper in the first place; nobody who had been a trapper would willingly, Kit thought, have given up his freedom for a stack of dollars. How deal with the deserters?

Even traders and clerks returned to the settlements reluctantly, he knew. There was William Bent, whose wealth already rivaled that of the Astors. Bent preferred to spend his summers with his wagon-trains on the trail, his winters in a Cheyenne tepee trading robes with the Indians. Gregg, the rich Santa Fe trader, expressed the same preference for the wilds. And Kit had heard English sportsmen, noblemen at that, insist that no man, even the most polished and civilized, who had once savored the sweet liberty of the plains and mountains, ever went back to the monotony of the settlements without lasting regrets and everlasting determination to return. All the men whom Kit loved and respected despised the settlements, loathed civilization, loved the wilderness. What could he think of runaways who stole the goods of such men and ran back to loaf in the cities? . . . Lousy skunks! Wait till Kit Carson lays hands on you!

Kit and his avenging comrade rode hard for the Arkansas.

When they got there, they found the *cache* empty, the furs gone, only a gaping, caving hole where their wealth had been! Not far off were four broken-down horses — Gaunt's finest animals — run to death — ruined. And sign showed that the two thieves had loaded the furs into Gaunt's old dugout canoe and set out down-river.

That showed what fools they were. They could not hope to navigate the Arkansas River by night, because of the sandbars everywhere. By day their canoe would be visible from every hilltop along the whole length of that almost treeless stream. The Injuns would get them, sure — if Kit did not get them first. It would mean hard riding to overtake them.

But Kit did not ride far. The Indians surrounded his camp, and he and his comrade forted in one of Gaunt's deserted log huts. There they remained, closely hemmed in, for several weeks. So great was their danger that Kit dared not remain outside his little fort even under the most pressing necessity without his rifle across his knees.

Meanwhile, Gaunt's men were playing in ill luck everywhere. When they came looking for Kit and the furs, hardly hoping to find him alive, they told him that their partisan had moved his headquarters to the Bayou Salade, the South Park (as it is called to-day) of Colorado. This Bayou Salade, or Salt Valley, was a favorite resort of the Indians and trappers and game animals, and consequently dangerous ground. Utes and Arapahoes swarmed there, cutting each other's throats. Buffalo — especially in winter — abounded there. Deer, bighorns, goats (as the mountain men called the pronghorn antelopes) were even more numerous. The streams were full of otter, beaver, muskrat. Nowhere in the mountains could be found a more alluring region — or one more dangerous.

On the way in, Kit's party was attacked by Indians, once, twice, and all the loose animals were stampeded and lost. Kit himself came near losing his hair that day.

Riding out to look for beaver sign, Kit and three companions saw four Indians on the trail ahead. Kit yelled, 'Come on, boys! Thar's the devils as stole our horses!' He struck spurs into his mount, and the other three galloped after him — straight for the thieving Arapahoes. The four Indians, seeing them charge, turned and raced away, with Kit close on their ponies' heels. He was getting ready to strike a *coup*. But suddenly, rising from either side of the trail, half a hundred warriors appeared in his path. Ambushed!

Twelve or fifteen to one! Kit had put his foot into the biggest kind of trap.

Kit never wavered, never hesitated. His one chance was to charge, to break through, to make a quick getaway. Boldly he rushed the line of naked, painted savages, facing their arrows, bullets, lances. His knees, his horse's shoulders touched the Indians as he plunged through. After him rode his three comrades, hell-bent for camp. They all knew better than to empty their rifles against so many red men. Had they fired, the Indians would have charged and killed them before their flintlocks could have been reloaded. It was a near shave, but once through the Indian line, they kept the savages at a distance with the threat of their rifles, and so got away. It was a near thing for them all.

In camp they found a wounded man, victim of the same Indians. Kit saw that the redskins were thick as grass on the prairie and were out for hair. He rigged a litter for the wounded man, swung it between two mules, and made tracks for Gaunt's main camp. From the Bayou they hurried on to the Middle Park and northward.

But the country had already been trapped over that season. They took nothing, and Gaunt was plainly busted. His men scattered, each small party trapping on its own hook. Kit and Rube Herring and Markhead decided to stick together. They did very well, keeping to the mountain streams and away from the prairies where, at that season, the Indians would be fighting and running meat. By the time cold weather set in, the three were headed for Taos with galore of beaver.

Kit had succeeded where Gaunt had failed. He had more beaver in his packs than he could have earned in years as a hired trapper. It came to him that he might as well scout round for some other steady man, throw in with him, and be a partisan of trappers himself. Why not? He had been poor a long time. Now — if only he saved his money — he might be a rich man. That day he vowed, as many a sailor has vowed before his ship made port, as many a cowboy has vowed on the long trail, that *this* time he would save his dollars.

Near the Raton Pass Kit overtook another party of trappers, and rode along with them with the biggest kind of heart. He had never been so rich, never felt so peart and sassy — not since he was knee-high to a beaver kitten. His high spirits hardly kept step with his new resolve to save his dollars. . . . And we have an accurate picture of what happened when he reached the Spanish settlements. . . .

CHAPTER VII
FANDANGO...

WITH Kit rode Rube Herring, six foot two in his moccasins, with long arms ending in hands of tremendous grasp, straight, wiry limbs, and any quantity of straight, black hair hanging over his brawny shoulders. By his side, swaying to the canter of a glass-eyed pinto, rode Dick Wooton, two inches taller than Rube, and as straight and strong as the barrel of his long rifle, a lad quick to fight and with an eye for pretty *muchachas*. Markhead, bold and independent as a hog on ice, sat his flop-eared mule with ease. And La Bonté, tall and bold as any of them, bestrode a fine Mexican mule nearly covered from sight by the enormous skirts of the Spanish saddle. After them trailed their pack-mules, weary under the heavy packs of furs.

'They struck the Taos valley settlement of Arroyo Hondo, and pushed on at once to the village of Fernandez — sometimes called Taos. As the dashing band clattered through the village, the dark eyes of the reboso-wrapped *muchachas* peered from the doors of the adobe houses, each mouth armed with a cigarrito, which was at intervals removed to allow utterance to the salutation to each hunter as he trotted past of *Adios Americanos* — "Welcome to Fernandez!" and then they hurried off to prepare for the fandango, which invariably followed the advent of the mountaineers. The men, however, seemed scarcely so well pleased; but leaned sulkily against the walls, their sarapes turned over their left shoulder, and concealing the lower part of the face, the hand appearing from its upper folds only to remove the eternal cigarro from their lips. They, from under their broad-brimmed sombreros, scowled with little affection upon the stalwart hunters, who clattered past them, scarcely deigning to glance at the sullen *pelados*, but paying incomprehensible compliments to the buxom wenches who smiled at them from the doors. Thus exchanging salutations, they rode up to the house of an old mountaineer, who had long been settled here with a New Mexican wife, and who was the recognized entertainer of the hunters when they visited Taos valley, receiving in exchange such peltry as they brought with them.

'No sooner was it known that *Los Americanos* had arrived, than nearly all the householders of Fernandez presented themselves to offer the use of their "salas" for the fandango which invariably celebrated their arrival. This was always a profitable event; for as the mountaineers were generally pretty well "flush" of cash when on their "spree," and as open-handed as an Indian could wish, the sale of whiskey, with which they regaled all comers, produced a handsome return to the fortunate individual whose room was selected for the fandango. On this occasion the sala of the Alcalde Don Cornelio Vigil was selected and put in order; a general invitation was distributed; and all the dusky beauties of Fernandez were soon engaged in arraying themselves for the fête. Off came the coats of dirt and "*alegría*" which had bedaubed their faces since the last "funcion," leaving their cheeks clear and clean. Water was profusely used, and their *cuerpos* were doubtless astonished by the unusual lavation. Their long black hair was washed and combed, plastered behind their ears, and plaited into a long queue, which hung down their backs. *Enaguas* of gaudy color (red most affected) were donned, fastened round the waist with ornamented belts, and above this a snow-white *camisita* of fine linen was the only covering, allowing a prodigal display of their charms. Gold and silver ornaments, of antiquated pattern, decorate their ears and necks; and massive crosses of the precious metals, wrought from the gold or silver of their own placeres, hang pendant on their breasts. The *enagua* or petticoat, reaching about halfway between the knee and ankle, displays their well-turned limbs, destitute of stockings, and their tiny feet, thrust into quaint little shoes (*zapatitos*) of Cinderellan dimensions. Thus equipped, with the reboso drawn over their heads and faces, out of the folds of which their brilliant eyes flash like lightning, and each pretty mouth armed with its cigarrito, they coquettishly enter the fandango. Here, at one end of a long room are seated the musicians, their instruments being a species of guitar, called *heaca*, a *bandolin*, and an Indian drum, called *tombé* — one of each. Round the room groups of New Mexicans lounge, wrapped in the eternal sarape, and smoking of course, scowling with jealous eyes at the more favored mountaineers. These, divested of their hunting-coats of buckskins, appear in their bran-new shirts of gaudy calico, and close-fitting buckskin pantaloons, with long fringes down the outward seam from the hip to the ankle; with moccasins, ornamented with bright beads and porcupine quills. Each, round his waist, wears his mountain belt and scalp-knife, ominous of the company he is in, and some have pistols sticking in their belt.

'The dances — save the mark! — are without form or figure, at least those in which the white hunters sport the "fantastic toe." Seizing his partner round the waist with the gripe of a grisly bear, each mountaineer whirls and twirls, jumps and stamps; introduces Indian steps used in the "scalp" or "buffalo" dances, whooping occasionally with unearthly cry, and then subsiding into the jerking step, raising each foot alternately from the ground, so much in vogue in Indian ballets. The hunters have the floor all to themselves. The Mexicans have no chance in such physical force dancing; and if a dancing *pelado* (greaser) steps into the ring, a lead-like thump from a galloping mountaineer quickly sends him sprawling, with the considerate remark — "Quit, you darned Spaniard! You caint 'shine' in this crowd."

'During a lull, *guagés* (cask-shaped gourds) filled with whiskey go the rounds — offered and seldom refused by the ladies — sturdily quaffed by the mountaineers, and freely swallowed by the *pelados*, who drown their jealousy and envious hate of their entertainers in potent *aguardiente*. Now, as the *guagés* are oft refilled and as often drained, and as night advances, so do the spirits of the mountaineers become more boisterous, while their attentions to their partners become warmer — the jealousy of the natives waxes hotter thereat — and they begin to show symptoms of resenting the endearments which the mountaineers bestow upon their wives and sweethearts. And now, when the room is filled to crowding — with two hundred people swearing, drinking, dancing, and shouting

— the half-dozen Americans monopolizing the fair, to the evident disadvantage of at least threescore scowling *pelados*, it happens that one of these, maddened by whiskey and the green-eyed monster, suddenly seizes a fair one from the waist-encircling arm of a mountaineer, and pulls her from her partner.

'Wagh! — La Bonté — it is he — stands erect as a pillar for a moment, then raises his hand to his mouth, and gives a ringing war-whoop — jumps upon the rash *pelado*, seizes him by the body as if he were a child, lifts him over his head, and dashes him with the force of a giant against the wall.

'The war, long threatened, has commenced; twenty Mexicans draw their knives and rush upon La Bonté, who stands his ground, and sweeps them down with his ponderous fist, one after another as they throng around him. "Howgh-owgh-owgh-owgh-h!" the well-known war-whoop, bursts from the throats of his companions, and on they rush to the rescue. The women scream, and block the door in their eagerness to escape; and thus the Mexicans are compelled to stand their ground and fight. Knives glitter in the light, and quick thrusts are given and parried. In the center of the room the whites stand shoulder to shoulder — covering the floor with Mexicans by their stalwart blows; but the odds are fearful against them, and other assailants crowd up to supply the place of those who fall.

'The alarm being given by the shrieking women, re-enforcements of *pelados* rushed to the scene of action, but could not enter the room, which was already full. The odds began to tell against the mountaineers, when Kit Carson's quick eye caught sight of a high stool or stone, supported by three long heavy legs. In a moment he had cleared his way to this, and in another the three legs were broken off and in the hands of himself, Dick Wooton, and La Bonté. Sweeping them round their heads, down came the heavy weapons amongst the Mexicans with wonderful effect. At this the mountaineers gave a hearty whoop, and charged the wavering enemy with such resistless vigor, that they gave way and bolted through the door, leaving the floor strewed with wounded, many most dangerously; for, as may be imagined, a thrust from the keen scalp-knife by a mountaineer was no baby blow, and seldom failed to strike home — up to the "Green River" on the blade.

'The field being won, the whites too beat a quick retreat to the house where they were domiciled, and where they had left their rifles. Without their trusty weapons they felt indeed unarmed; and not knowing how the affair just over would be followed up, lost no time in making preparations for defense. However, after great blustering on the part of the Prefecto, who, accompanied by a *posse comitatus* of "Greasers," proceeded to the house and demanded the surrender of all concerned in the affair — which proposition was received with a yell of derision — the business was compounded by the mountaineers promising to give sundry dollars to the friends of two of the Mexicans, who died during the night of their wounds, and to pay for a certain amount of masses to be said for the repose of their souls in purgatory. Thus the affair blew over; but for several days the mountaineers never showed themselves in the streets of Fernandez without their rifles on their shoulders, and refrained from attending fandangos for the present, and until the excitement had cooled down....'[1]

Kit was always fond of poker, pony-running, and dancing. Apparently he took this sort of thing rather lightly in those days, as his own version of the matter simply says that he 'disposed of beaver for a good sum, and everything of mountain life was forgotten for the time present.'[2] Peters moralizes on this to the effect that Kit had now 'become impressed with the highly important fact that there existed a much wiser course to be pursued' than such riotous living. If Kit had any dollars left in his possibles that autumn, it must have been owing more to the suspension of gayety due to this fracas than to any self-restraint he imposed upon himself. For he soon joined Captain Lee, one of Bent's agents, in an expedition to the mountains.

[1] Ruxton, *In the Old West*, pp. 288-95. Quoted by courtesy of The Macmillan Company. Of this passage, Ruxton says, 'The Mexican fandango *is true to the letter. . . . This is positive fact*.' The italics are Ruxton's. This adventure of Kit's has been ignored by earlier biographers, though vouched for by a gentleman and an officer in H.M. 89th Regiment, and printed in *Blackwood's Magazine*, 1848, under the title, *In the Far West*.

[2] Grant, *Kit Carson's Own Story of His Life*, p. 30.

CHAPTER VIII
BLACKFOOT SCRAPE

LEE'S party followed the old Spanish Trail through the timber and mesas of the Apache country, passing prehistoric ruins perched in the walls of deserted canyons — on into the range of the friendly Utes. To White River, the Green, the Uintah. There, at the forks, they found Antoine Robidoux, thin and sardonic, established in a substantial trading post. Lee halted to dispose of the goods he had brought so far on muleback, and his men went into winter quarters with the men of the fort.

Kit was lodged in the tepee of Blackfoot Smith, one of Bent's best traders. Blackfoot Smith was a lively man, fond of singing sentimental ballads, fond of good liquor, fond of gambling, never without a squaw. If no more courageous than a trader ought to be, he was at least very competent as a trader. In his lodge Kit lost some of his own taciturnity.

Trading was a leisurely proceeding in the mountains. Kit would sit with Smith and his comrades in the warm tepee, playing poker all day long, or it might be *monte*, seven-up, euchre — it was all the same to Smith. Outside on the snow they would hear the crunch of moccasins, and a solemn Indian head would appear in the door. Then the horse-hide cards would be laid aside, and Smith would take out of his packs a Green River knife, some Galena pig-lead, or a three-point Nor'west blanket, red as blood. The Indian would sit and consider, finger the blanket and haggle for half an hour, confer with his Indian companions, all the while inhaling Smith's good honey-dew tobacco, and at last take his buffalo robes and go away without the thing he had come to buy. All this time Smith would be busy with the sign language, with smatterings of Ute or Comanche, with Spanish and trapper talk, displaying his wares to the best advantage and using no end of salesmanship. But when the customer lifted the door-flap and was gone, Smith went cheerfully back to his cards, knowing that the Indian would return, talk things over again, and at last buy. Slow sales did not matter; the profits were enormous. And even if the Indian did buy, the bargain was not concluded. He might change his mind and bring the blanket back next day — next week. What could the trader do but acquiesce? He could not afford to make enemies.

Smith was fond of giving advice, and, taking advantage of his place as host of his snug tepee, often enlarged upon the advantages of having a woman and a lodge. The trapper — that amateur Indian — had much to learn from the redskins, and one of the first lessons was that an Indian wife made an excellent helpmeet. This lesson Smith tried to l'arn Kit:

'*Valgame Dios!* Your shanty is a pore make-out compared to this hyar Injun lodge, Kit. Leaky and cold and open to the weather, and whar's the fire when you come in at night half froze for a hot kettle of soup? And your fingers too cold to strike a light. Why should you freeze all winter like a starvin' coyote? Your rifle shoots plumb-center; she makes 'em come; you kin throw plenty of fat cow, and you know whar to lay hands on a pack of beaver when you want it. It's time you womaned, Kit, and that's a fact.

'Maybe you're thinkin' of some sickly gal from the settlements, thin as a rail and pale as a ghost, pretty as a pitcher and so fofurraw she's good for nothin'. Maybeso you've sot yore eyes on some wench to Taos or Santy Fee. Do you hear now? Leave the Spanish slut to her greasers and the pale-face gal to them as knows no better. Put out and trap a squaw, and the sooner the better.

'What a mountain man wants is an Injun woman — one who can pack a mule, make meat, dress robes, make moccasins, cook, pitch a lodge, ride all day and then give birth to a likely young 'un after sundown. Look at me, Kit. I'm warm, I'm comfortable, I'm happy as a bear in winter quarters, with the old gal settin' hyar beside me. When I come home at night, froze stiff with cold and starvin', I kin see the big yaller lodge all lit up like a lantern among the pines, and I know when I go in, thar the old gal will be, with a good fire burnin' and the kettle steamin'. And before I kin get out of my wet moccasins and peel off my coat, my woman will have the warm water ready for my feet and a bowl of coffee under my nose. Then I kin set and smoke my pipe and listen to the lonesome wolves a-howlin' on the hills and the wind roarin' through the pine-trees. Do you hear now? My old gal is *some*, she is. I wouldn't swap her for all the beaver in Bent's big lodge. She kin make a home for me wherever grass grows. And you kin lay to that!'

Kit was more conscious of the disadvantages of mating. He observed that the wife's relatives would be something of a problem, as they swarmed into his lodge, expecting entertainment. But Smith pointed out that the difficulty could be obviated if Kit would take a woman from a tribe at war with the Indians where he made his home, for then the relatives would never dare to call. Of course, he admitted, his squaw liked the ribbons and hawk's-bells and scarlet cloth in the trader's packs as well as any woman likes fofurraw, but — he bragged — he could always check her extravagance with half a length of lodge-pole, if she got too fractious.

'Kit, air you goin' to trap a squaw?'

'Not while I can set in *your* lodge, Blackfoot.'

For indeed, the big lodge was very comfortable. A fire no bigger than you could put in your hat heated the taper tepee evenly and without smoke, and all drafts were turned upward by the sheets of lodge-cloth stretched along the walls halfway up from the beds on the ground. And Kit knew that, when camp moved, Smith's woman would have the tent packed and the mules on their way in less time than it would take to smoke a pipe.

Sitting in Smith's lodge, Kit heard of the big battle in Pierre's Hole, lasting all day — how fifty Blackfoot bested four hundred white men, killed a partisan, and then made their getaway in the night. He heard, too, of how Broken Hand had played hide-and-seek with the Blackfoot for days — how he had starved and lost horse and beaver and even his rifle, and how, when he did get back to camp, his long hair was white as snow and nobody there knew who he was. White Head they called him now, Kit heard.

One morning Robidoux missed six horses — his best stock — and, on calling the roll of his men, found a California Indian missing. Antoine had no intention of losing his finest animals, and looked around for some one to send after them. Antoine had more than a score of trappers in his employ, and there were Captain Lee's men to choose from, provided Lee gave his consent. But from them all Antoine, that wise old fox of the mountains, asked Kit to go. Lee made no objection, and Kit rode over to the Ute village and arranged for a celebrated Indian trailer to go along.

Antoine gave them the best mounts he had left, and they set out at once. For a hundred miles the trail was plain. Then the Ute's horse gave out, and Kit went on alone. Thirty miles farther along, Kit caught sight of the runaway. The Indian, his white blanket coat outlined against the dark mass of the horses he was herding before him, was going at a steady lope. He kept a close watch of the back trail, and saw Kit almost as soon as Kit saw him. Both men stripped the covers from their rifles, and the Indian turned his herd toward the shelter of some timber not far ahead on his left.

Kit looked to the priming of his rifle and put spurs to his horse. The ground was so uneven and the pace so swift that Kit had no good chance to fire until the Indian reached the timber. He knew that, if he missed, the Indian might kill him from cover before he could reload — or even afterward. Kit reserved his fire, and rode for the Indian for all he was worth.

The Indian saw who was after him, and rode hard, leaving the stolen horses to go where they pleased. He reached the timber, halted, raised his rifle to fire. Evidently the Indian was full of confidence, thought himself safe now.

But before he could pull the trigger, Kit's rifle cracked. The white blanket coat swayed from the saddle, the Indian gun went off harmlessly into the air, and the bay horse trotted off riderless into the timber. It looked as though the Indian was gone beaver before he hit the ground. Kit reloaded and rode forward to investigate. He took no chances with a dead Injun; like as not the man was shammin'. Yet he was pretty sure. This ought to count for *coup* number six.

The Indian was thrown cold, a bullet-hole over his left eye. Kit got down, pulled out his skinning-knife, jerked off the scalp, slapped it against the dead man to rid it of blood, and tucked the trophy under his belt. Old Antoine would want to see that. Then Kit rounded up the stolen horses and set out for the fort. He traveled slowly, in order to allow the worn-out animals to recover. Four days later he drove them into the corral, handed over the scalp to Antoine, and received the reward he had been promised.

Soon after, Lee had a chance to sell out to White Head Fitzpatrick, and Kit was out of a job. He hired with White Head, but finding too many trappers in the camp, decided to hunt on his own hook. He did very well, for he kept his men in the mountains — away from the dreaded Indians of the Plains.

But in the mountains he met a foe more dangerous than the warriors of the prairie.

One day Kit set out to kill some meat for the camp. He found sign of a band of elk, approached them carefully, took aim, and dropped a fine bull. No sooner had the elk fallen than Kit heard the rush of something behind him. He turned, and a glance showed two huge grizzlies almost on top of him. Their snouts were raised to display the terrible fangs below; their long claws raked the ground as they came charging. He could not fire, for his gun was unloaded, and in any case he could not have killed both with one bullet.

His knife could make no impression on their shaggy coats, their tough hides underlaid with thick fat. Kit knew that he must escape by running, and he knew, too, that a grizzly can run as fast as a good horse for a hundred yards or so. These were not fifteen yards away when he saw them.

Instantly, Kit threw down his gun and ran — ran for a small thicket of aspens, the nearest trees — ran his fastest. He reached the trees, chose the thickest trunk, clambered up, swinging himself upon a bough that bent dangerously under his weight. One of the bears tore off his moccasin with a sweep of its claws as he drew up his legs into the swaying, limber tree. The little leaves were all a-tremble, and so was Kit.

He looked down into the gaping maws of the two grizzlies. They were furious. In those days the silver-tip had not learned to fear man, and even to-day he will savagely attack any one who comes upon him suddenly or unawares. Kit's shot had startled the bears, and at once they jumped him. Kit thought the two of them would outweigh him ten times. And they were determined to have him down.

Kit knew the full-grown grizzly could not climb far, but his tree was so small that climbing only a few feet would be enough to bring him within reach of their claws. Moreover, the weight of the bear would break the slender aspen, and he saw the first one coming up as far as it could.

Quick as thought, Kit broke off a branch, and wielding it as a club, beat the nose of the bear as hard and as fast as possible. The grizzly drew back with a roar of pain, and began to tear up the roots of the trees. Old Ephraim, as the trappers called the grizzly, was the smartest critter in the mountains, and the roots flew from those terrible paws, torn to shreds by the steel hooks at the end of them. It looked as though Kit could never hope to see another day. He felt gone beaver. He was as good as dead, he thought.

The old-time grizzly rubbed out many a trapper. The Indians counted *coup* on him as though he were a man, and few of them lived to tell of it. Back and forth swung the shaking aspen, with Kit clinging to its slender branches, while the two bears growled and tore at the trunk and roots. Then the elk enticed one of them away, and half an hour later the second gave it up and lumbered off into the hills.

Kit waited a long while, then slipped down, picked up his rifle, and hurried back to camp. The trappers went hungry that night, unless they had beaver meat — no dainty dish — in their traps. But Kit was satisfied to be alive and kicking.

Soon after he and his party joined old Gabe Bridger on Green River. Two hundred trappers made rendezvous there in two big camps with more than their own number of camp-keepers, squaws, and children. Their partisans' supplies had not come from St. Louis, and they had to buy from traders at outrageous prices, paying two dollars for a pint of sugar or coffee or Du Pont powder. And the commonest kind of American-made blanket (and in those days all good blankets were imported) cost twenty-five dollars. Kit left rendezvous poor as a bull at the end of the rutting season. His 'beaver' was all in the pockets of the traders. He had to start at once on his fall hunt. With fifty others, he set out to the country of the Blackfoot.

The Blackfoot, and especially the Piegans, were extremely hostile to invaders of their country. In 1831 they had made a treaty with Kenneth Mackenzie at Fort Union on the Missouri, and had agreed to trade with him alone. Even Mackenzie's trappers were not allowed to work in their country, and it is small wonder that others were driven out. Kit and his comrades were butting their heads against a stone wall in their new enterprise. But who can blame them? For years they had slaved at their difficult trade, and not a man of them could answer satisfactorily the question in his mind, 'Whar's the beaver as ought to be in my possibles?' The traders had it all. 'Twas all to taverns and to wenches. Well, the Blackfoot range was rich with fur. They would go after it, get rich.

On Big Snake River the Blackfoot ran off eighteen head of horses one night. Kit was leader in the pursuit, taking with him eleven men. After pushing over the snow for fifty miles, he came up with the Indians, owing to the fact that they had to break trail through the drifts for the stolen animals. When the trappers appeared, the Indians drove their horses up on a hillside where there was little snow. They themselves easily kept out of reach of the trappers' rifles, because they had snowshoes and the trappers floundered through the drifts, up to their armpits in snow.

Kit thought he would be lucky to recover his mounts, and signaled with a blanket for a parley. The Blackfoot, as usual, were quite ready to have a good look at their enemies before the fight began. They sent a brave forward to sit and talk.

Kit went forward to meet him, and the two talked between the lines. Kit roundly accused the Blackfoot of thievery, and demanded the return of his animals. The Indian blandly insisted that his party were quite innocent. He said they had been raiding the Snakes, and never dreamed that the horses belonged to the white men. They had no desire to fight the trappers, he said.

'Well, then,' Kit urged, 'why not sit and smoke? Lay down your arms. Talk it all over peacefully.' The Indian agreed, and signaled back to his men in the pines. Kit shouted to his trappers on the hillside.

Both parties posted a guard over their weapons and came forward to make talk together. They sat down in two lines facing each other. The trappers were sulky and angry, the Blackfoot bland and smiling. The chief offered his pipe, and all the men smoked it in turn very amicably. Then the palaver began.

Kit kept insisting upon the return of the stolen animals. The Blackfoot, who outnumbered the whites three to one, felt that was asking a good deal, after the trouble they had had in stealing them. Moreover, possession was ten points of the law in the Old West, and the Indians had no idea that the trappers could forcibly take back their animals. Neither had Kit, though he bluffed them for all he was worth.... At last they sent and brought in five broken-down wolf-bitten pack-horses, worn out by the hard push through the snow. The rest of the herd was in plain view from the place where they were all sitting.

The chief refused to do more than that. The white men had no right in his country. His thirty-five braves sat and smiled insolently at the white men.

Kit jumped to his feet. No Indian ever affronted him with impunity. 'Come on, boys; get your rifles!' he yelled, and led the way back to the stacked arms on the run.

The Indians also dashed back to their own weapons, making better time on their snowshoes than the moccasined trappers could do. Then the shooting began, and there was a very lively skirmish. The trappers found that Mackenzie had armed his Blackfoot well, and things got pretty hot. Three to one was not considered odds ordinarily, but this was a different proposition. The trappers charged the Indians, and both parties fought from behind the cover of the trees and rocks at close range.

Markhead, daring and reckless as ever, exposed himself in the front of the fight, Kit loading and firing at his side. Both were counting *coups* right smart, when suddenly Kit saw his target — an Indian's head — exposed from behind a boulder. Taking aim, he was about to shoot, when he saw another savage aiming at Markhead. To save his friend, Kit shifted the barrel of his rifle, fired, and killed the Indian. At the same moment, Markhead fired, and the two men were unprotected, their single-shot muzzle-loaders empty.

Kit's antagonist let out a yell, jumped forward, and covered Kit, knowing that the white man could not escape. Kit dodged back and forth, trying to avoid the aim of the Blackfoot. But Mackenzie's rifles shot plumb-center, and, in spite of Kit's efforts, the Indian's bullet found its mark. Kit felt the jolt and the pang in his neck and shoulder. His right arm hung useless, and blood gushed freely over the fringes of his hunting-coat.

When Kit was hit, the trappers retreated to the place where they had left their horses, and the Blackfoot followed them gladly. The trappers took cover among some trees on a little knoll and dressed Kit's wound. One of them slapped a bunch of beaver fur upon it, and bound it up with a strap of soft buckskin cut from the skirts of his coat. It was sundown, and the night came on bitterly cold. Not daring to light a fire, the trappers lay unprotected all through the long dark hours, rifle in hand.

Kit had only his saddle-blanket. He was shaken and chilled from shock and loss of blood. And the bandage failed to stop the bleeding. He passed a miserable night, hardly hoping to live until daylight. But the terrible cold increased, and at last froze the blood as it flowed from his wound, and so stanched it. The cruel frost saved his life. But the cure was almost worse than the injury.

Next morning they found the Blackfoot still cheerfully waiting for another battle. But Kit and his men had had enough. They started back to camp with the five miserable animals the Indians had given them.

As soon as they arrived, Bridger and a large force of men set out to punish the Indians, but could never overtake them. For the Blackfoot, finding that the white men would not fight, took their ponies home — to dance and count *coups* by their comfortable lodge-fires.

Once more Kit Carson was impressed by the fighting qualities of the Plains Indian. His experience was to stand him in good stead when, in after years, he was given the duty of subduing the wild tribes of the Southern Plains.

But now he gladly turned his face to the mountains again. His party rode down into the valley of Green River for the summer rendezvous.

CHAPTER IX
CAPTAIN KIT CARSON

DURING these years in the mountains, Kit Carson found many things to do besides trapping. The fur was valueless in summer, and in winter the beaver slept in his lodge, safe within his thick walls of frozen mud and intertangled logs — a barrier as impregnable as reënforced concrete. Half the year the trapper could not work at his profession. Improvident and generally in debt, he had to find something to do to maintain himself. Sometimes he traded butcher knives or alcohol to the Indians on the Platte or the Arkansas. Sometimes he hired as hunter to some trading fort, and made endless journeys in quest of meat to feed the *engagés*. Sometimes he was packer, guide, or rode out to capture wild horses. If he was a married man with a lodge to maintain, he had to visit the mountains for tepee poles at one season, and the plains for buffalo hides to make the covering of his lodge at another. Like enough, his wife would want to see her relatives sometimes, or there would be a gathering of trappers at Taos, Santa Fe, or Bent's Fort, which he would not want to miss. All these enterprises kept him on the move, kept him busy all the year round. Trapping occupied him only in the autumn and in spring: from October or late September until the streams froze, and again from the thaw in April until the beginning of summer.

The trapper, too, was a touchy individual, hard to boss, and was apt to leave an unpopular *bourgeois* at a moment's notice, as soon as his debt for equipment was settled. The true-bred mountain man could not endure interference, even of the most benevolent kind. An old trapper once advised a greenhorn to mind his own business on the prairies, never so much as informing a friend of a lost pack or a strayed animal. 'Leave him alone; he'll find it out. If you tell him about it, all you'll get is a cussin' for your pains. He thinks he can take care of himself.' Headstrong, fierce, and uncertain as so many Indians, the mountain men could be loyal only to those they considered real men; self-interest played little part in their plans. It is probable that, but for the debts of these men to their partisans, few of the bands could have been held together through one season.

And so we find, in the records of travelers of those days, that Kit is forever turning up at the most unexpected times and places, times and places never mentioned in the brief notes on his life which he dictated but could not read. All that other mountain men did, he did. For he was not only one of the best of the trappers — a master in his profession. He could make a saddle, mend a gun, build a fort, make snowshoes or canoes or bullboats, run bullets, handle an axe or a knife, dress skins, kill and butcher his game. He was an expert packer, cook, wrangler, trailer, teamster, breaker of horses. He knew the habits of big game animals, and was versed in the customs and mental processes of the Indian warrior. And in the course of his work he had formed the habit of mapping in his mind the significant features and lay of the land drained by the streams whose waters he followed looking for beaver.

In addition to this skill, he had won the reputation

of absolute truthfulness and reliability. In the wilds of the mountains, truth was not a matter of conveying impressions. Truth had to be literal, objective, factual: life and death might depend on a misconception. If a man made one false statement, his comrades, his Indian customers, would never trust him again. The clear eyes and clear head, the unquestioning blue eyes of a man who first made sure he was right and then went ahead, these gave Kit Carson the character which made him the power he was among the trappers. No doubt his gifts as a linguist helped him here. For though Kit was unable to read, he readily mastered the languages of people who came in his way. He spoke Spanish as the Mexicans spoke it, French as the French Canadians knew it, Cheyenne, Ute, and Comanche, besides smatterings of other Indian tongues. He was swift in the use of the racy dialect of the mountaineer. It was only when he attempted correct English that his lack of education and book l'arnin' betrayed him.

To these linguistic talents he added, in time, a pretty command of the sign talk or gesture language universal among the Plains tribes.

All this skill and the caution it engendered made his courage irresistible when he attacked, and he was a crack shot. He had realized the world he lived in more accurately than other men, and successful action was therefore easier for him. He was a man of some imagination, a man who made mistakes, but who seldom repeated them. Constantly in his memoirs he uses the expression 'concluded to charge them, done so,' all in one sentence. To Kit decision and action were but two steps in one process.

Small wonder that his fame was already widespread in the mountains. Among the fraternity of trappers he was loved, feared, admired. Inevitably, such a man was marked for a partisan, a leader of mountain men. All he lacked was the money to finance his improvident followers. And Kit, seeing this, began to take thought for its accumulation. He saw that he must save his beaver.

An expedition to the desert provided a stake. Returning to New Mexico, his party struck the Rio Grande del Norte, entered San Luis Park, heading back to the good old Bayou Salade. There Kit made up his own band. He and Joe Meek, Bill Mitchell, and three Delaware Indians — Jonas, Tom Hill, Manhead — went off on their own hook to make their hunt south of the Arkansas River on the small streams which flow from the mountains there.

These six men formed the nucleus of Kit's band, that celebrated organization known as the 'Carson Men.' Later it grew and grew in numbers, and for many years Kit had as many as forty or fifty men in his employ. The full story of the adventures of those men would make another Book of the Round Table. But as yet the band was small, and at the very beginning Kit found his generalship put to the test.

They were now in the range of the Comanche Indians. Bill Mitchell always hankered to hunt in that region, which he knew very well, for Bill had once been a warrior in that tribe — by his own account, a very famous one. He had heard of a gold mine in the Comanche country, and, thinking he might find it and get rich, he packed his plunder on his mule, rode into their camps one fine day, took a Comanche woman, and began to look round for the gold.

Legends of gold in the Wichita Mountains still persist, and the bones of more than one party of prospectors — some Spanish, some American — lie in the canyons there. One of these legends deceived Bill Mitchell. But at last, failing to find the gold, he became discouraged. Then the problem was to leave his Comanche friends without offending them. Bill was stumped for a while. But finally he allowed his horses to stray away. Then, making the plausible excuse that he must go after them, he mounted his old saddle-mule, rode over the hill, and pushed hard for the nearest fort — Bent's big lodge on the Arkansas. That ended his life with the Comanches, but he was always eager to return to their country. He acted as guide to Kit's party.

It was new country to Joe Meek, that happy-go-lucky wit and wag from Virginia, one year younger than Kit, with the long nose, full eye, and small, thin-lipped mouth which betrayed his lively temper. He was the youngest, though not a man in the band was over twenty-five. One of the Delawares — Jonas — was not yet out of his teens, though he handled his long rifle with the best of the trappers.

The Delawares were typical members of their tribe, which had taken over the white frontiersmen's weapons and mode of life, differing from them only in speech and blood. Wherever the white man went, there would be found a Delaware acting as guide or hunter. They were well-armed, and combined the skill of the white hunter with the sure instinct of the red man. The Plains Indians hated them bitterly because they killed the buffalo in their country, and more than once tried to exterminate a band of Delawares who had ventured upon the prairies. But bow and lance could not match those riflemen, and the Plains Indians were badly worsted. The Delawares were cleanly Indians, proud and industrious in their hard profession, and the mountain men freely accepted them on equal terms. This fact is significant, for the mountain men despised the Spaniards, the Mexicans, the French Canadians, the greenhorns from the settlements, even the soldiers of the Regular Army. Kit knew what he was about when he chose the three Delawares for his band.

One spring morning the six trappers were riding across the bare prairies, heading south. Not a tree was in sight, not a bush. The mountains far to the west showed dimly, blue and vague in the sunlight. The plains undulated gently away and away, one rolling wave like another, far as they could see. But Bill Mitchell knew his way. He rode steadily forward, his red gee-string flying in the wind, his bare buttocks pounding the Spanish saddle.

Suddenly Bill reined up, and Kit, looking where he pointed, saw a round black dot on the hilltop ahead. 'Injuns!' It was the season for war parties. The trappers halted and stared at the Indian scout's dark head, waiting to see what it portended. They had not long to wait.

For the Indians, hidden behind the hill, were at once informed by their scout what had happened. The white men had halted: it was clear that they had seen the Indian scout. Over the hill they came pell-mell — mounted on their best horses — racing to count their *coups*.

All at once the skyline sprouted lances, tossing like

grass-blades in the sun, then black-and-white eagle-feather crests, horses' heads, naked, painted warriors. The charge was on. At the same moment the war-whoop, like the quick chatter of a machine-gun, pulsated upon Kit's ears. The whole hillside was covered with Indians.

'Comanches!' yelled Bill Mitchell, and looked to Kit for orders.

'Two hundred of 'em, or I'm a nigger!' said Joe Meek.

The Comanches were magnificently mounted. They always were. They had more horses and better horses than any Indians on the plains, and they 'ate and slept horseback.' They constantly raided the vast herds of Spanish horses on the *haciendas* to the south of their range — the best animals on the prairies. Kit knew he could not run away from them — and there was no cover within miles. Six to two hundred!

'Fort, boys!' he sang out, and jumped off his mule, jerking out his scalp-knife before his moccasins touched the ground. The mule, with all a mule's instinctive fear of Indians, tried to break away, almost jerking the stocky little man off his feet. But Kit caught the lariat close to the animal's head, and, as it reared back, passed the keen edge of his knife across its taut throat. He jumped clear. While the mule staggered, coughing out its life, drenching the short grass with blood, Kit snatched the cover from his rifle, looked to the priming, glanced round at his men.

They had followed his example. Already three mules were down. Hastily, Kit and his comrades flung themselves prone behind the kicking carcasses, pointing the muzzles of their rifles toward the coming warriors. The ground shook with the beat of eight hundred hooves, the sunlight glittered on the long, keen lance-points, and lit up the garish war-paint upon the naked bodies, the flares and blotches of color upon the spotted ponies. Feathers streamed from lance and war-bonnet. On they came. It was magnificent, and it was war. Kit yelled a warning.

'Bill, don't shoot yit. Hold on, Joe! Let the Delawares shoot fust!' Joe and Bill nodded, grim-lipped, never taking their eyes off the charging Indians. It was hard to lie idle, finger crooked on trigger. But they knew Kit was right: it would never do to empty all their guns at once. Three shots against two hundred savages!

Kit was gesturing swiftly to the Delawares. 'You killum, *sabe?*' And Tom Hill, muttering a word to his red companions, grinned knowingly, at the same time drawing a bead across his dying mule on the foremost warrior. Tom looked very strong and competent, lying there, his long body covered to the knees with his straight, black, unplaited hair.

Already the horses were so near that Kit could see the whites of their excited eyes. Ahead rode the chief, his lance wrapped with shining otter fur, his war-bonnet streaming behind. *Crack!* The three long rifles spoke together. The chief tumbled from his saddle, struck the ground on his head just in front of the little barricade, and was dragged away by his frightened horse, having tied his body to the end of the lariat. The charge split, and swept by in a thunder of hooves, the rush of crowding horses, white smoke in clouds from the rifles, a rain of arrows lancing the dust.

Immediately the redskins turned and charged **again,**

and this time Kit and Bill and Joe swung round, faced the other way to meet them, aimed and fired as steadily as though they had been armed with repeating rifles or machine guns. Again the charge was split, and the Indians dashed by. Two were left on the ground. Bill let out a war-whoop.

But now, whirling round in a moment, the Indians raced back. The white men's guns were empty, they knew. The Delawares had not had time to reload. Now they could ride the whites down, lance them with impunity — out of reach of their sharp knives. Back they came, whooping and laughing with expectation of an easy victory. One of them recognized Bill Mitchell and called out, taunting the white man, as he came: 'Lean Bull, your hair is mine. Now I am going to make the ground bloody where you lie!' In a flash the Comanches were upon them.

But the Indians never reached the whites. They could not force their ponies to approach the dead mules. The smell of the blood drove their horses crazy, and the charge ended in a *mêlée* of bucking, rearing animals, circling round the trappers, too unruly to allow their masters to draw bow and shoot. Kit's stratagem had saved his band.

And now Jonas was taking aim; his rifle blazed, and the laugh on the face of the taunter changed as he swayed from his saddle and toppled to the ground. The Comanches saw him fall, saw Manhead aiming, saw Tom Hill strike the butt of his rifle on the ground, too much in a hurry to ram down the charge. They turned, they retreated, and the frightened ponies made the retreat a rout. The trappers stood and cheered.

But the Comanches, in spite of their losses, could not

believe that six men could stand off two hundred. Armed only with bows and lances, they re-formed and charged bravely, only to split and retire to the hilltop again. All the while, whenever there was a lull in the fighting, the trappers were busy with knife and hatchet, deepening their defenses, and at last they had an adequate fort even against arrows at short range. And as often as the Indians charged, they kept them off, firing in shifts, so that always three rifles were loaded.

Again the Comanches charged and retired. Then their medicine man, shaking a big rattle, confident of his power to turn bullets, led them on. Kit dropped him, and the redskins, finding their medicine no good that day, sat and smoked and talked things over on the hill. During the council, the Comanche women came down to carry off the dead and wounded.

It was scorching hot in the midday sun. The trappers had no shade, no water. Their throats were parched with the heat, the fever of excitement, the dust, the reek of the rifles. Flies swarmed about the dead mules, and stung fiercely. And the women cursed and scolded, shaking their fists helplessly at the whites, threatening vengeance: 'Dog-faces, I throw filth at you. Cowards! Women! Wait till the council is over! I shall dance over your scalp!'

What wonder if Manhead, following the custom of his people, raised his rifle to throw the squaw who screamed out insults? It was an easy shot. And to kill a woman under the eyes of her men was always rated a brave deed — a *coup* — by the Indians.

Kit saw what Manhead was doing, and commanded him to hold his fire. Kit was no Davy Crockett, to have a hand in the slaughter of women.[1] Many a squaw died at the hands of trappers, and even more fell to the guns of greenhorns and men in uniform. But Kit Carson had only scorn for the skunk who would shoot a woman, red or white. Manhead let the squaw go.

All day the Indians sat by and the trappers stood them off. The fighting became rather half-hearted — young men galloping round and round in circles, displaying their marvelous horsemanship, shooting from under their ponies' necks, sometimes dashing up close to the dead mules to throw an arrow into them in sheer bravado. The trappers killed a few horses, but could do little damage to the young men. For every Comanche had a loop of hair rope braided into his pony's mane, and swinging in this, with one heel on the animal's back, was able to screen himself entirely behind his racing horse.

Come night, the trappers were still waiting, hungry, dry, anxious, their shoulders sore from the kick of their guns. They lay low and watched the ragged silhouette on the skyline. At last it melted away. For some time they remained in their fort. Then, thinking the redskins had gone, they got to their feet, stretched arms and legs, moved about. Most of the charges had come from the hill to the south. The bodies of the mules on that side of the fort were so thick with arrows that Kit could not lay the flat of his hand anywhere upon them without touching an Indian shaft. Yet not one of the trappers was seriously hurt. Meek, letting his love of a good yarn smother his better judgment, estimated forty casualties on the

[1] *Vide* Davy Crockett's *Autobiography*.

Indian side. Kit hammered two brass tacks into the stock of his rifle.

The mountain men had to leave their traps, their possible sacks, their costly saddles, their well-filled packs. Slinging their blankets over their shoulders, and carrying only their hatchets, knives, and rifles, they sneaked away through the prairie starlight. After a mile or two they settled down into a steady dog-trot which they maintained all night. Back to the mountains. Back to camp. Bill Mitchell told them it was nearly eighty miles to the nearest water.

It was all in the day's work. Trapping was a business — a profitable business at its best — when not interfered with by war. And war — wal, war allus was a pore make-out. You might save yore skin, but whar's yore mule?

CHAPTER X

THE ARAPAHO GIRL

THE rendezvous on Green River that summer of 1835 was a lively one. Green River was admirably adapted to the great fair of the trappers. There was good water, there was wood, there was grass in plenty for their animals, and game abounded thereabout. The valley was easily accessible to the pack-trains of the traders who came to buy the furs of the mountain men. And the Green was equally convenient for the Indians who brought their buffalo robes and horses to trade to the white men.

That summer the traders arrived early, and the trappers, having bought their new equipment, could spend the balance of their beaver in carousing. So heavy were their packs that season that the rendezvous continued all the summer until time for the fall hunt, and the traders disposed of all their goods and alcohol galore.

The trappers, having camp-keepers to do all the work about camp for them, lived like lords, feasting on boss and hump-ribs, Injun dog and panther, indulging their luxurious taste for *coffee* — with *sugar* in it — and treating each other to the atrocious whiskey brought from the settlements in flat kegs made to fit the back of a mule.

Only one thing was lacking to the rendezvous — and that was the fandango. No *señoritas* ever went so far from their adobe New Mexican towns, and as for white women — white buffalo would have caused less excitement in the valley of the Green!

But the mountain men, not to be deprived of their fun, contrived to fix up fandangos with the Injun gals. For a little way downstream from the camps of the various companies of trappers was a large encampment of Arapaho Indians, who had come to trade, to stare, to break the monotony of their lives by as many a spree as they could pay for with robe or peltry.

The Arapahoes were a sociable people, fond of a good time, rather more affable, genial, and sensuous than other tribes. Social dances were held nightly in their camps. Some of these dances were for men, others for women, but the trappers preferred the Soup Dance to all the rest.

The Soup Dance was a dance of coquetry, and both men and women took part. The dancers stood in two lines facing each other, the young men on one side, the girls on the other. Near one end of the line stood half a dozen musicians round the drum. Between the two lines were placed two or three kettles of appetizing soup, and every girl was armed with one of those polished black spoons made of the horn of a buffalo or — it might be — a huge yellow translucent spoon of mountain sheep's horn.

When the music struck up, each girl in turn selected some man in the line opposite, and advancing and retreating in time to the music, she approached nearer and nearer to her partner, extending the spoon filled with soup toward his lips. The young man, muffled in his blanket, as soon as he saw that he was chosen, was expected to dance forward in pursuit of the girl until she had returned to her own place in the line. Then he retreated in time to the music, and she followed in turn, still offering the spoon. This backing and filling went on amid the laughter and gibes of the onlookers until the girl finally gave in and allowed her partner to drink from the spoon. Immediately after, she ran, and he pursued. Having overtaken her, he might fold his blanket about her, and they would stand and talk for as long as she chose, in plain sight of the camp, but enveloped in the privacy of the young man's blanket. The girl always expected a present from her partner next day in recognition of her favor.

If the young man did not wish to dance with the girl who chose him, she was likely to become offended and throw the soup in his face. Again, if she allowed him to drink the soup, but did not wish to talk to him, he was supposed not to persist, but to let her get away. In that case the girl did not expect him to give her a present.

The Soup Dance among the Arapahoes was merry, but decorous. After the trappers got through with it, it was much less decorous and very much merrier. For the white men immediately attempted to discredit the spoon of soup. They knew better ways of spooning than that. Kisses took the place of soup — and kissing was a complete novelty to the Arapahoes, who up to that time had greeted each other by rubbing noses.

The Arapaho girls took to kissing readily enough, and so far as can be learned the custom of rubbing noses has gone completely out of fashion among them.

When Kit arrived at rendezvous, he found the dances in full swing, for Captain Drips and certain other partisans had had their bands of trappers on the ground for some time.

The Arapaho Girl 117

Having disposed of his furs and bought his new outfit, Kit went down to the Arapaho camp one night to see the fun. The moon was full, and a light breeze rustled the leaves of the big cottonwoods along the river. A mile downstream he could see the big yellow lodges of the Arapahoes, tall taper lanterns, firelit, and could hear the thumping of the tom-tom and the high voice of the crier summoning the dancers. Passing through the ponies snuffling over the dusty grass, and beating off the savage Indian dogs which attacked him as he advanced, Kit came at length to the fire of logs where the dancers were already at it.

He found that the trappers had filled the places of the Arapaho young men, and every one was dancing at the same time, prancing and whooping and swinging the gals in a manner totally strange to Indian ideas of propriety. The leader of them all was a gigantic Frenchman, loud and boastful, drunk as an owl, who shoved other men aside and danced with whatever young squaw pleased his blurred eyesight. This sort of thing caused resentment, and Kit asked who the bully was.

Shunar, or Shunan, they told him. A new man in Captain Drips's camp. A great fighter, six foot three in his moccasins, thick-set, and strong as a bull. He had already thrashed every man in Drips's band, and was rarin' to go against anybody who showed fight. '*Enfant de garce!* I'm de beeg buck of dis lick, *moi!*' he was accustomed to brag. Whenever he wished to change partners, he would seize the girl, push her partner away, and bellow, 'By Gar! I wan' dees woman. *Vous savez? Vamos!*' None of the trappers had the heart to oppose him.

That night he had appropriated a pretty girl, who seemed embarrassed by the way in which he embraced her. She was powerless to escape from him, and danced with her head down, giggling, and avoiding the kisses he tried to force upon her. When the drum stopped, she managed to break away, and fled precipitately down the river-bank. Shunar was not too drunk to follow. But the trappers had the laugh on him, for he came back presently, panting, alone. The girl had run away from him. Somehow Kit was relieved at that. But Shunar was sulky and quarrelsome as long as he stayed at the dance.

Next day Kit saw the girl riding her white pony about the camps in company with her girl chum. She was certainly very pretty. Her glossy hair shone bright and blue as a rifle-barrel in the sun. Her lively, laughing face was touched with vermillion at the temples. She had fine teeth, and her rounded forearms were banded with silver. Her soft, doeskin skirt, decorated with patterns of red paint and colored quill-work, was fastened by a broad belt bossed with incised silver disks, and its long fringes hung down about skin-tight leggins and little moccasins that fitted like gloves. She rode astride, with a robe about her hips, rode with grace and ease, proud of the handsome high-pommeled saddle, the white pony with its Spanish bit.

In those days the Arapaho women had not been taught to be ashamed of themselves, and in warm weather they often went about with only a skirt on, so that their bodies were bare from the waist up. Kit's eyes followed her wherever she went. What lithe beauty, what red-gold skin, what firm little breasts!

Who was she? Kit made inquiries. Her name, he

learned, was Waa-nibe, Grass Singing. Her father went by the name of Niekahochithinaahnie, which meant Running Around or Running in a Circle. Kit knew it must refer to that custom of the Indian scouts, who, when they discovered enemies from a hilltop, always ran or rode in a circle on the hillside, as a signal to their comrades. A warrior's name, that. And Niekahochithinaahnie was a member of the Lime Crazy Society. The name and the membership marked him as a man of standing in the tribe. Kit found that the father of Waa-nibe was a man in middle age. He had a large, clean tepee decorated with pendants of buffalo dew-claws and bearing a painted likeness of the Thunder Bird. This lodge, Kit was told, would insure fair weather as long as it stood in the camp. Niekahochithinaahnie was manifestly *some* in Arapaho circles. Kit made a friend of the Indian with a present of powder and ball.

It began to look as if Kit might trap a squaw. Perhaps he dreamed of her: Kit believed in dreams. At any rate, Kit made the Soup Dance every night. She was always there, but she took a perverse delight in dancing with everybody else. The choice of a partner lay with her, and there was nothing much that Kit could do about it. Arapaho convention did not tolerate conversations between young men and women, and Kit had no desire to outrage the family of the girl he was thinking of. His one chance lay in a kiss and a conference after the dance.

At last it seemed that he might win her favor. As the drum struck up, she filled her big yellow spoon with soup, and stood facing him, smiling, while her chum whispered in her ear. Kit saw that they were talking about him, and he waited to see what would happen. It was about time she paid him some attention, he thought.

Just then Shunar forced his way into the line shoulder to shoulder with Kit, and both men waited for Waa-nibe to make her choice. Kit did not yield an inch, and the two men, conscious of their rivalry, their bodies pressing against each other, stood there, while the drum beat and the thin Indian voices rose in the song.

Waa-nibe came slowly forward, holding out the spoon. Each man thought she was choosing him: Shunar, because she was coming toward him; Kit because she looked at him. Shunar did not wait to touch the spoon. He let out a whoop, stepped forward, and made a pass at the girl, intending to embrace her. But Waa-nibe, stepping back, lowered her spoon, and flung the contents straight into his face. Then, laughing, she threw down the spoon, and put up her lips for Kit's salute. Kit concluded to kiss her, done so. Then she ran for the shadows of the cottonwoods along the river, and Kit was after her like a flash. He found her ready for huggin'.

Shunar had flung up his hands, swearing, his eyes shut, his beard dripping. The trappers stood around, shouting with mirth at the bully's plight. But when Shunar got his eyes open and glared around, their guffaws died on their lips. For a moment he stood there. Then he went sulkily off, mounted his pony, and rode into the brush. The talk and laughter broke out again: it would be a long time before the camps heard the last of that. 'The Injun gal's full of the old ginger, and you can lay to that!'

Next morning, Waa-nibe appeared riding a spotted pony — one of Kit's band. Shunar got drunk — drunker than usual — after he saw that. He knew what people were saying — behind his back: that Kit had him bested. *That* little runt!

Waa-nibe did not attend the dance that night, nor the next. Kit hung around, hoping she would come. But she did not. Among the tepees he saw her. But when she saw him, she ducked into a near-by tent, and he could not be sure which one. He went back to rendezvous. Next day he rode over to talk with her father in the big painted tepee.

Kit sat and made talk in the sign language, and Niekahochithinaahnie sat and smoked and nodded, nodded understanding, uttering an occasional *Hnh*, *Hnh*, until Kit had finished, had made the sign Cut Off.

Then the father talked in turn in that same language, the poetry of motion, briefly and indifferently, and began to fill his pipe again. It appeared that the final word in the matter would have to be that of the girl's brother. That was the custom, said the Indian.

And who was her brother? Where was he?

Detenin, Short Man. Maybeso he would be coming home pretty soon. Maybeso not.... Maybeso! It seemed to Kit that Niekahochithinaahnie, who had been so cordial before, was suddenly indifferent, not to say unfriendly. Why, Kit had no idea.

After a time Detenin pushed through the doorway of the lodge. When he saw the white man, he straightened suddenly, paused, and would have gone away. But his father spoke, and the young man came in, sat down. He paid no attention to the white man. The boy's father went on explaining why Kit was there in the lodge. But Detenin, his dark face bent sourly above his naked chest, never gave Kit a glance from his hostile eyes. His heart was hot against the whites. He would not talk with Kit.

After a long time, Kit got at the core of the difficulty. It was that Shunar, that Bad Medicine, that Frenchman. He had followed Waa-nibe into the timber, had found her alone, had caught her, had held her in spite of all her struggles and all her cries. He had told her that she had thrown soup in his face. Now he would have his revenge. The more she struggled, the more the big man laughed.

And so Detenin, who had heard of the matter only that morning, was unable to talk to a white man. He hated them all. He would shortly go upon the warpath and kill all he could.

But what of Waa-nibe?

Oh, Waa-nibe, like other modest girls of her tribe, always wore the hair rope, the Indian belt of chastity, under her dress when away from the camp. The guardian rope, passed round her waist, was knotted cunningly in front. Passing between her thighs, one end was wound round either leg, and fastened at the knee. Shunar had had everything his own way until he encountered the rope. Then he had to pause, to get out his knife to cut it, and Waa-nibe, tearing her hands free, drew her own blade, stabbed him, and made her escape. No, she had not hurt him much: she was in a hurry to get away. She had kept it all a secret until to-day: then she had told her brother. Detenin's heart was hot against Shunar — against all men with hats. He would not listen to Kit, he said: his ears were closed.

Kit did not wait to talk.

In the trapper camp, Kit found Shunar fighting drunk. Already the big bully had beaten up two of Kit's comrades. They lay in the ditch, their faces a bloody jelly. Shunar saw Kit coming, and at once began to brag louder than ever. He could lick any Frenchman, Dutchman, Spaniard, or Injun in de whole worl', by Gar! As for *sacrés Américains*, if one of dem skonks look at *heem*, he would cut a stick and switch heem. . . . Kit knew very well why the man was so quarrelsome. And he was happy to see it.

Kit walked under the big Frenchman's nose. His cold blue eyes shone up into the giant's ruddy face: 'Hyar's an American, Shunar, and the meanest kind at that. Thar's plenty of men in camp can lick the hindsights offen you, only some of 'em are scairt of your brag. Wal, you caint scare me. Keep your trap shut. I won't take sich talk from no man. If ye don't, I'll rip your guts out! Savvy?'

Shunar had just what he wanted. Without answering, he turned on his heel, hurried back to his lodge. There he mounted his best horse, and, armed with his rifle, rode up and down the grassy flats along the river, cursing and bragging that he could lick all the Americans on earth. Now he would rub Kit out. Afterward, if he chose, he would have the Arapaho girl.

Kit ran to his shanty, not far off, and, fumbling among the packs, snatched up the first arms he found there — a pistol. He was furious, and so full of fight that for once he forgot his caution. His best buffalo horse was tethered near. Hopping on its bare back,

Kit galloped away to find his enemy. He had only the one shot in the pistol.

Duels between mountain men were invariably to the death. They began shooting as nearly at the same time as possible, and kept on until one — or both — had gone under. Neither Shunar nor Kit had any notion of any other end to that conflict. One or the other must die the death.

Shunar saw Kit coming, and rode to meet him, his rifle laid along his thigh, ready to shoot off-hand. Kit, rash and impetuous, rode straight up to him, and his pony shouldered the Frenchman's a little aside. The two men looked into each other's faces. Shunar must have seen the red rays about the pupils of Kit's eyes, so close they were. Perhaps those red rays shook his confidence a little. Perhaps not.

'I reckon I'm the man you're gunning for?' Kit drawled, his soft voice full of menace.

Shunar could hardly answer in the affirmative. Kit had pushed up so close that the rifle was no longer pointing just where Shunar wished. Kit, armed with his pistol, had all the advantage of position at the moment. Shunar played for time.

'No,' he said, shifting the rifle in his hand ever so little, so as to cover Kit. Kit saw what he was up to. But Kit was willing to give the man his chance. He could see the Frenchman's trigger finger. For a moment neither stirred. Kit knew he could not miss at that distance.

Bang! Both fired together. Kit swayed sidewise from the saddle. The ball passed his head, cutting his hair, and the powder burned his cheek. But Shunar had dropped his rifle. Kit's bullet had shattered his wrist, passed through his upper arm as well. Shunar was helpless. He began to beg for his life.

Kit wheeled his pony back toward the shanty — to get his other pistol. The trappers on the side-lines were happy at the outcome of the fight. That night Kit hammered another brass tack into the stock of his rifle. The boys crowded round to buy the drinks.

Once more Kit sat in the lodge with Detenin and his father. And once more he made talk in the silent language of signs. The two Indians no longer showed animosity. Waa-nibe was called into the lodge, and, as she sat by Kit's side, her father threw a blanket over the two of them. Then he went outside to display the new gun, the three mules, and the five Nor'west blue blankets which Kit had given him. Firing the gun into the air, he made public announcement of the marriage, and loudly invited his relatives to come to the feast.

Soon after, they came, and among them provided Waa-nibe with a new tepee and everything necessary for her new home. Then Kit hurried back to get his traps from his shanty in the trapper camp. He was in haste to get them before his friends heard that he had womaned.

The trappers were loud in their congratulations on the duel. But to all such felicitations, Kit replied apologetically. He was a little ashamed of himself — about the second pistol. 'Shucks, boys, I ought to ha' throwed him the *fust* shot!'

This is the true story of Kit Carson's duel with Shunar, as I have got it from the Arapaho Indians in Oklahoma. There have been many versions, most of which have followed Peters, the authorized biographer, who assumes that Shunar survived. Peters based his account upon Kit's own memoirs, which make no such claim, leaving the matter in doubt. Sabin [1] bears witness to the killing, and also to Kit's undying hatred for Shunar.

Kit says nothing about the Arapaho girl. At first glance, this does not seem to bear out his character for perfect honesty. But when we look into the circumstances of the dictation of the memoirs, the imputation cannot stand. When the memoirs were dictated, Kit was an officer of the United States, a famous personage, and married to a lady of good family in Taos. He had several children. Squaw-men were then unpopular, as the Indian wars were in full swing. Peters, the authorized biographer, was an ass, and Kit was undoubtedly aware of the fact. Wisely, Kit decided to let those memories of the good old days rest unprofaned, unpublished. His caution (during his lifetime) was perfectly natural and right.

Inquiry among the Arapaho Indians in Oklahoma brings to light the names and facts of this story, and also the name of Waa-nibe's sister, Hisethe, Good Woman, who, in the normal course of events, by Arapaho custom should have been Kit's second wife. The mother of these girls was an Atsena, a woman of the Gros Ventres of the Prairie, a tribe split off from the Arapaho and speaking the same language. In Kit's lifetime, the Atsena were allied with the Blackfoot, and this may account for the fact that Waa-nibe has sometimes been referred to as Kit's 'Blackfoot woman.'

Kit was twenty-six when he married her. Watan,

[1] *Kit Carson Days*, pp. 166 and 632.

whose grandfather was related by marriage to Detenin, Kit's Arapaho brother-in-law, told me that Waa-nibe was considered a good girl by her people — good to look at and a good housewife. And she was all of this, by Kit Carson's own account, often repeated. She made him a good home.

The Arapahoes were the most artistic of the Southern tribes. They were good singers, and had many songs. Their language was full of broad vowels, soft liquids, and smooth diphthongs, so that Indians of other tribes preferred to sing Arapaho songs, even though they could not understand the words. The Arapahoes had a more elaborate art than the other Indians, and their lodges, clothing, furnishings were generally neater and more colorful. Their mythology was voluminous, and they were much given to long rituals and ceremonies. In short, in the person of Waa-nibe, Kit had come into contact with a culture far more beautiful and poetic than anything the settlements in Missouri or New Mexico could at that time show. It was his first experience of an ordered society.

What was the effect of all this upon Kit Carson?

Well, his wife was named Waa-nibe, Grass Singing. And he called her Alice! . . .

Kanathehahade, Coming Horseback, an Arapaho chief and brother-friend of Niekahochithinaahnie, had a camp of his people fifteen miles below the rendezvous at this time. It may be that Kit took his bride down there and away from the trappers, who would be sure to bring out all the pans and kettles in camp to celebrate his wedding with a civilized shivaree. Owing to the very limited opportunities for young folks to become acquainted before marriage, the Arapahoes, like other Plains Indians, made it a custom for the groom to refrain from his bride for several weeks after the wedding. The two newly-weds slept together in the same bed — or rather spent the night talking there, and the girl-wife wore about her body the protective rope — the Indian belt of chastity — until the period of courtship was over. In fact, during all this time the young men friends of the groom made his lodge a sort of club-house, and slept there nightly in large numbers.

It is not known whether Kit was aware of these customs, or whether he observed them, if he was. The trapper generally was not very patient with the conventions of other people — Indian, Spaniard, or greenhorn — as the case might be. Probably Kit moved his new lodge right into the camp of his comrades and began his married life at once.

CHAPTER XI
THE PRETTIEST FIGHT

On leaving Green River, Kit and his men made their fall hunt on the Yellowstone and Bighorn, the North Fork of the Missouri, the Big Snake. Kit led the way, riding at the head of the cavalcade, proud of his handsome squaw, his picked trappers, his string of pack-mules, his big yellow lodge — even though he rarely looked round at them.

They found Thomas McKay, a Hudson's Bay Company partisan, and traveled with him to Fort Hall. It was a starving march. They found no game, and had to live on such roots as they could find. When these failed, the men would bleed their horses a little, catch the blood, cook it, and eat. Hard doin's when it comes to that! The country of the Humboldt Sink was no place for a trapper. But at last they found Indians, bought a fat horse, and filled themselves up again.

After a rest at Fort Hall, they went on the hunt for buffalo, which they found within four days' travel from the fort. Hunting buffalo was no matter of an afternoon's walk. When Kit first saw the plains, in 1826, buffalo were everywhere. But now, ten years later, the great herds were much diminished, and often no bison would be seen for months at a time near the fort. Buffalo hunters had to strike out and keep going in the likeliest direction, and *keep going* until they found meat.

Coming back with the buffalo meat, Kit encamped outside the fort in his lodge. The Blackfoot, who had followed the hunters back from the buffalo range, played a pretty trick upon them one fine morning.

The Indians were very bold and cunning on that occasion. They waited until almost morning. Then, in the gray dawn, two of them walked up to the corral gate, let down the bars, and slowly drove out all the animals — right under the nose of the sleepy horse-guard, who thought they were two of Kit's party driving their stock to water. It was some time before the loss was reported. Then there were curses, loud, deep, and heartfelt. Every man in the fort and camp was left afoot. They could not pursue. Once more the Blackfoot scored an easy victory. The trappers knew by the moccasin tracks that the thieves were Blackfoot, for every tribe made shoes of a different model.

There was Kit, and there was Waa-nibe, afoot, four days' ride from meat, and no way to get to it. They had a handsome lodge, and no means of moving it. Kit's grudge against the Blackfoot was getting so severe that he had a hard time keeping it under. But this time he had to hold in. There was nothing to do but sit down and wait.

At last McKay turned up again, and Kit bought new horses from him. But Waa-nibe sighed. Gone was her fine white pony — gone the spotted pony Kit had given her. And she had to struggle with strange, half-broken pack-mules, quite unlike the familiar old nags her brothers had given her. Together they moved into rendezvous on Green River, making camp near the mouth of Horse Creek.

That year Kit had lived in a comfort to which he had never been accustomed. Now he had always a warm, tight lodge for home, and plenty of moccasins, garments, and a soft bed supported on willow-rod mats to rest upon when he came back from hunting. His rifle provided Waa-nibe with all the hides and meat and sinew she needed for her housekeeping, and his traps paid for all the fofurraw any woman might desire to make herself and her lodge attractive. Beads, ribbons, hawk's-bells; paint to make the face pretty and protect the skin from sun and wind; brass kettles and Spanish bridles: he could easily supply them all.

Kit was happy with Waa-nibe, who — if Watan is right — was a faithful, competent, jolly wife. On Horse Creek she presented Kit with a baby girl.

The child arrived in a little tent set up to one side, and presided over by the mother's aunt or grandmother. Certainly Kit was not there, for no modest Indian woman would have permitted a man — not even her husband — to witness the birth of her child. But afterward, when the child was heard crying, Kit might go and take a look at the little pale copper body flushed with red, at the bright eyes and red-gold hair. Then he felt that he was, sure enough, some punkins.

Then the grandmother washed the baby, packed it on a baby-board, swaddled in the down of cat-tail rushes, and called in Waa-nibe's women relatives and friends, who came bringing the presents they had prepared in anticipation of the little fellow's coming.

A year of contact with Arapaho culture may have softened Kit Carson's fierceness a little. But it had not much improved his taste. He called that baby girl Adeline. Sweet Adeline!

From Green River, Kit rode with Fontenelle to the Yellowstone, the Musselshell, the Big Horn, Powder River. Near by their fort on Powder River the Crow

Indians camped in their lodges — the most beautiful and spacious of all the tepees on the plains, lodges trailing painted streamers from the tip of every pole.

The Crows told Fontenelle that the Blackfoot had been swept away by smallpox. For the trappers, after their bitter experiences in the Blackfoot country, had come prepared to fight, and could not understand why it was that no one attacked them.

That plague of 1837 was one of the most terrible in history. These Indians had never had the disease before, and had not developed immunity. As a result, whole tribes were swept away.

Kit and his comrades were not sorry. Now they could hunt in peace, they thought. As they moved through the country, they often caught sight of deserted Blackfoot camps, where whole bands had been rubbed out by the plague. The tepees stood, silent and smokeless, the dead lay unburied. Wolves ran about the village, fat and impudent, and when the men rode near, flocks of noisy ravens rose fluttering above the conical tents, like burned paper whirling in the smoke above a fire. The Indian dead hung in swarms in the dead trees in which they had been buried, and brown buzzards sat in rows along the bluffs, gorged with human flesh, drunk with ptomaines. Occasionally they would see bodies at the foot of these bluffs, where the last survivors had flung themselves down in despair. But the trappers took good care not to go too near those noisome cities of the dead.

That winter the snow was deep. The horses could not graze, and had to be fed on cottonwood bark. And then the buffalo, starving for forage, came hooking and goring the horses, driving them from the boughs the trappers had cut. Kit had to build fires near his horses to save them from the bulls. But with only buffalo to combat, the mountain men felt easy enough. They passed the time pleasantly.

But on the North Fork of the Missouri, Fontenelle's men found that the Blackfoot were not exterminated, after all. They cared little for that, however. They had expected to have a battle or two, and had come prepared, several bands united, and every man well-armed. It was the biggest band ever brought together in the West up to that time, barring the gatherings at rendezvous. The trappers felt that it would be rather a pity to leave the Blackfoot without a lesson. Their packs were full. Kit had taken almost five hundred beaver: six packs, six hundred pounds! At six dollars the plew!

One day, as he advanced, leading the van, he ran into the Indians. They were also on the march, strung out over the prairie, moving camp. Kit rode back to warn the trappers of the main body. A quick council was held, and, leaving ten trappers with the fifty camp-keepers, the remaining forty loaded their rifles, mounted, and followed Kit back to the place where he had seen the Injuns.

There they were, strung out across the grass: pack-mules, travois, dogs and all, and the trappers gave a whoop and charged them.

The Indians were at a disadvantage, having to defend their women and children, and being surprised at that. But they outnumbered the trappers, and made a brave retreat, riding always between the whites and the fleeing women, keeping them back, standing them off as best they could, losing a dozen men or more.

For two or three hours the trappers followed, shooting and charging, happy at last in having things their own way.

It was a slow, running fight. But at last the Blackfoot women had got enough of a start to allow their men to leave them. The warriors turned back in force and attacked savagely. Then the trappers found out what a wonderfully different critter an Injun charging is from an Injun charged. The Blackfoot rode through and through the trappers' lines, breaking them up into small parties, each of which put up the best fight possible, every man running or shooting as discretion dictated.

Sometimes one side had the advantage, sometimes the other, but the Indians had the trappers going — and going fast — straight back where they had come from.

Kit and Cotton Mansfield rode together. Kit happened to be in the lead when Cotton's horse fell, pinning him to the ground. Mansfield lost his rifle in the fall; he could not draw his leg from under the horse; he thought his time had come. Six braves raced to count the *coup*, scalp-knives in hand. Mansfield yelled after Kit:

'Tell old Gabe that old Cotton is gone!'

Kit heard the call, wheeled, jumped from his saddle. Raising his rifle, he fired, throwing the foremost Blackfoot cold. The other five turned and made a dash for cover, and Cotton had time to rouse his horse, regain his feet and his rifle, mount, and ride. Other trappers saw the little fracas, and only two of the five Indians saved themselves. Those mountain men! They made 'em come!

Meanwhile, Kit's horse became frightened and left him afoot on the prairie, with all those Indians charging through the lines. Now his gun was empty, and the Blackfoot saw it. On they came, whooping and racing — each eager to be the first to strike the helpless trapper. Kit looked around for help. But Cotton had gone!

Just ahead of the Indians a man named White was riding for his life. Kit called to him, and the man swerved aside, passed near Kit, offered his hand. Kit jumped up behind him, and the horse rushed on, carrying them both to safety.

Doc Newell shot an Indian. He dismounted to scalp him, and in his hurry to get the hair, entangled his fingers in the gun-screws with which the Indian had adorned his braids. When the sharp knife pricked his head, the savage came to, and for about three minutes Doc would have sold his life for one of the gun-screws and called it a bargain. He could not free his hand. The Indian had two hands free, and a knife in one of them. And his friends were coming to help him!

But Kit shot the Indian with a pistol, and stood off the rest. Doc got away.

The Blackfoot, too, had their rescues. A squaw, whose horse was shot from under her, came near being captured. But catching hold of the tail of her husband's horse, she was snatched away from danger fast as the animal could run.

At camp, the trappers made their stand. Now they had a hundred men, and the Indians, they thought, would give in. But the Blackfoot had had so much fighting that day, they cared for nothing. They took cover behind a pile of rocks a few rods away and began to shoot. The trappers fired too, but could do no damage. So, seeing the Indians were asking for it, the trappers decided to charge, and charged at once. Having come to the pile of rocks, the trappers fought in the good old backwoods fashion, from rock to rock and tree to tree. Says Kit, 'It was the prettiest fight I ever saw, ... I would often see a white man on one side and an Indian on the other side of a rock, not ten feet apart, each dodging and trying to get the first shot.'[1]

At last the trappers pushed the Indians so hard that the redskins fell back, and ran, having lost several scalps. Not one of the trappers was killed, nor any very seriously wounded. Through every account of this fight runs the evident joy of battle of the sassy mountain men. 'It was the prettiest fight I ever saw!'

It was just that. Just the kind of dashing battle for a pretty story of Western romance — scenario by Robert Louis Stevenson — with all manner of loyal comradeship and heroic rescues, and all the casualties where they should be — on the Indian side!

Oh, the good old days! No wonder Kit, in after years, looked back to his time in the mountains as the happiest period of his life. Those *were* the days!

Ah, those early Indian fights — before the Indian had acquired really good weapons and the skill to use them! As jolly and exciting and almost as harmless as a football game.

But before Kit died, he was to find that times had changed. Adobe Walls, Fort Phil Kearny, Platte Bridge, the Little Bighorn ... what a different story Indian wars told then!

Well, it was good while it lasted.

[1] Grant, *Kit Carson's Own Story of His Life*, p. 41.

CHAPTER XII
KIT CARSON'S LUCK

IN those days the trappers were a gay, rollicking, gallant lot of boys, most of them hardly out of their teens, few indeed more than thirty years old. In the mid-century and after, when the trapper got into literature, the mountain men were old, and it became the fashion to talk of 'old trappers.' But in the thirties they were all young, bold, and full of fight. The beaver trade, though near its end, was still flourishing. And they *would* go back to that rich and disastrous Blackfoot country.

What a range that tribe had! An ample, grassy, gently rolling plain, studded at long intervals by groups of hills or isolated buttes, well-watered by many streams flowing from the Rockies, and 'plumb lousy' with big game. Beaver swarmed there, choking the creeks with their dams. No wonder the mountain men could not resist going in.

But the Indians, too, had their temptations. *Coups* and scalps, but most of all, horses. Horses the Indian never could resist, even the horses of his friends and allies. And so there was an endless series of raids and thefts, battles and stampedes — all on account of the horse. Any warrior would stop fighting to capture a horse which had broken from his enemy. Horses were the Indian's wealth. Of course, he might take wild ones and break them, and often did. But it was much more thrilling to steal well-broken animals from his enemies. Hunting was drudgery; war was sport; but stealing horses was the business of his life.

How the white man hated him for it! The animosity of Westerners toward horsethieves persisted well into the cattle-range days, when the rustlers adopted the Indian's easy methods. Indeed this intense hatred for the horsethief is not yet entirely dead, though nobody thinks of lynching the man who steals a car worth a dozen head of cow ponies. What would the history of the West have been if Henry Ford had been born a hundred and fifty years earlier?

Kit's men fell in with the Blackfoot again. The trappers, forty in number, drove the redskins to an island in the Yellowstone, where they forted and stood off the white men until dark. Kit lost one man — a Delaware. Then they went back to camp.

Returning in the morning to finish the fight, they found the Indians gone. There was much spilt blood within their fort of logs, and a well-worn trail led to a big hole in the ice, into which the trappers believed the warriors had thrown their dead.

The Blackfoot village was not far off, and the white men knew that before long the Indians would return in force to revenge their comrades. Kit posted a lookout on a lofty butte near camp. About ten days later the lookout signaled *Injuns*. Then he hurried back to camp. Indians were coming, coming by the hundred, riding over the snow. The lookout estimated fifteen hundred warriors! Fifteen hundred to sixty. Sixty — counting the camp-keepers. But the trappers were not unready. During those ten days they had built a stone fort on the hill — a fort that only cannon could damage. They went into it with all their possessions, and coolly awaited the Blackfoot charge.

Lying in their fortification that night, they could

hear the songs and war-whoops of the redskins at their war-dance, and all night long the flat, regular thumping, as the warriors beat the parfleche which served as drum on the war-path. It was certain that in the morning the Blackfoot would attack.

They came, sure enough. But finding the fort so strong, and anticipating the terrible price they would have to pay for its capture, they contented themselves with firing a few random shots. It looked as though the cussed redskins were going away.

Then the trappers, full of fight, leaped upon their breastworks and taunted the Indians, making insulting gestures, cursing them in all the tongues — Spanish, French, English, and Indian dialects: Dog-faces! Cowards! Women! Buffalo calves! Puppies! *Sacré! Caramba!* God-dam skunks! Come and fight! Come on! We'll larn ye who this country belongs to!

But the Indians, not caring to commit suicide, rode calmly away and sat in council within sight of the fort. Then the trappers watched, waited, hoped for a fight after all, surmised what was being said as the long pipe passed from hand to hand among the chiefs in the circle on the snow.

The circle itself they could not see, for hundreds of young men sat on their horses about it, listening to the chiefs. All that Kit and his men could see was that dark mass of horses, topped with yellow-white buffalo robes, black dots where the heads showed above the robes, an occasional tuft of white and black where a war-bonnet crested the row.

Presently there was a stir. The Indians mounted. Half of them rode toward the country of the Crows; the rest went back the way they had come. The trappers cursed the Indians for cowards, and regretted their lost opportunity. No one molested them while they remained in that camp.

At rendezvous all was grumbling and despair. The fur trade had paid too well. Now everybody was at it, and the country was getting trapped out. Keen competition was proving the ruin of the industry. Alcohol — recently banned by the Government — was smuggled into the Indian country, and each company used it freely to debauch the red men, and so destroyed the next year's trading for robes. Trappers, too, were lured from their partisans by offers of better pay. To make matters worse, rival partisans slandered each other to the Indians, stooped to shady practices, even instigated robberies of their rivals at the hands of the red men. White Head had been so robbed by the Crows, and laid it at the door of the American Fur Company. Things were going to the dogs. Already old Gabe Bridger, White Head Fitzpatrick, Milton Sublette, Frapp and Gervais, had dissolved the Rocky Mountain Fur Company. And for some obscure reason, the price of beaver was going down.

But the country of the Blackfoot was still good hunting ground. Bridger organized a large party to invade it, and Kit took his band along. There they ran into Joseph Gale, builder of Fort Hall.

He had had no luck at all. His men were without animals, without ammunition, without traps, packs — almost naked. The Blackfoot had stripped them bare, had killed several, had wounded Dick Owens. Kit's band, which was in the lead of the main party, took Gale's helpless crew under its protection, and together they moved on.

One day Kit sent two trappers forward to look for beaver sign. Not long after he heard firing in the direction they had gone. It grew in volume, and rapidly came nearer. It was clear that the Blackfoot had seen his men, that they were driving them back.

Gale's men had no ammunition. They would be useless. Kit had only a few men. He knew from the volume of the firing that the Blackfoot were coming in large numbers, and fast. He quickly decided what he must do. Near by was a patch of dry brush, which fronted upon a grassy valley lying toward the sound of the firing. Kit told Gale's men to lead the horses into the brush, well back out of sight, and to hold them there. Kit cached his men in the edge of the brush, facing the open. There they waited, rifle in hand.

Suddenly two figures came sprinting into the open, running at top speed over the grass, jumping and dodging, zigzagging to avoid the bullets of their enemies behind. 'Thar's Meek, anyhow,' said Kit. 'They'll never take his hair.'

The two trappers raced on, burst into the dry brush with a snapping of brittle twigs, flung themselves down, while Kit's men fired. The Blackfoot ran boldly out into the open, but when the thicket of brush bloomed with bursts of smoke and spurts of fire, they hurried back into cover.

Meek, panting from his hard run, was talking: 'Blackfoot, Kit.... The biggest kind of band.... We run like rabbits.... Let me have some powder, will you?'

The Blackfoot remained under cover, not caring to charge across the open against so many crack shots. There they remained. Probably they might have gone away, but, hearing the squealing of a mule which had been hit, they could not. The horseflesh was too great a temptation. They kept on firing into the brush, swarming behind the rocks and trees just beyond the open valley. And as often as a bullet struck the twigs above the trappers' heads, leaves and twigs crumbled into dust, dry and brittle with the summer's drouth.

Kit was satisfied that he could stand them off until Bridger came. Old Gabe would not be long, even though the wind was wrong to carry to him the sound of the firing. Too bad. It was a stiff breeze — right in Kit's face. If it had only been blowing the other way, now. It seemed as though it would be merely a matter of an hour's siege. No fun at all.

All at once, Kit raised his body to a sitting posture, staring out of his cover toward the Indians. The other men, too, left their rifle pits, and a worried look came into their eyes. A moment before they had all felt peart, but now they began to have the smallest kind of heart for the battle.

Several Indians were running about over the grass, stopping, stooping, waving something red in their hands. And wherever they stopped, a thin curl of white smoke rose into the blue sky, plain against the side of the hill. Fire! They were firing the grass! And a stiff wind was blowing directly toward the patch of dry brush where the trappers had forted! Fire!

If it caught, they would have to run for their lives, and the red men would get them, sure. Better that than to burn in the brush. And even if the men did escape, the horses were sure to be taken. Kit for once had a problem he could not solve.

The trappers watched and waited. The columns of

white smoke spread into curtains, joined, and advanced. Already they could catch the first faint whiffs of burning grass; already the air was glazed with the heat, shutting out all clear vision of the Blackfoot. The black, charred ashes behind the fire grew and widened rapidly, and the little flames danced merrily forward, struck the longer grass in the bottom, swept forward, towering, racing, straight for the dry thicket, now clouded with the smoke.

Behind him, Kit could hear Gale's men cursing the restless horses, and the noise of their plunging and kicking. Some of them would be breaking away soon. But what could Kit do? He lay and watched the fire advance. Probably the Blackfoot would be following it up, ready to charge through the smoke when the trappers broke cover.

The flames died down a little as they climbed the slope to the brush. There the grass was shorter. But still the wall of flame came on, pushed by the wind, right up to the edge of the thicket. The first twigs caught and snapped, and Kit and his men sprang from their rifle pits, helpless, ready to run.

Kit had done all he could. The heat beat upon his face, the smoke choked him. He had done his best. Only Providence could save him now, he thought. But, with his usual determination, he stuck to his post. He wouldn't run until he had to.

But for some reason the fire burned out, died down, at the edge of the thicket. The men beat out a few crackling twigs with their coats. The brush did not catch fire. Slowly the smoke drifted away, the Blackfoot were seen still behind their rocks, the horses grew quiet. One animal only had run out. It was dropped by a stray bullet, and from behind it two trappers were pouring their bullets into the Indians on the rocks.

And then — 'The Injuns are running! Old Gabe is coming!'

Sure enough, the Indians were leaving. No more shots were heard. No more smoke bloomed from the rocks where they had been. They had failed to drive Kit out — his *medicine* was too strong. And their scouts brought word that the main party had heard the firing, that Bridger and his men were racing to join the scrap.

When Bridger arrived and heard the story, he looked disgusted. He himself was always getting into tight places, always getting the worst of it. He had been Kit's partisan — along with White Head — and he took a certain pride in Kit's prowess. But this sort of thing was a leetle too much. Old Gabe spat into the black ashes. 'Kit Carson's luck,' he said.

The party moved on to Stinking Creek.

From that time the superstitious trappers had a great respect for Kit's *medicine*. It was strong. Whenever he organized a party, he had no lack of candidates to choose from.

CHAPTER XIII
HELL'S FULL OF HIGH SILK HATS

AT rendezvous on Wind River, Kit and his companions were amazed to find encamped a nobleman and sportsman, Captain Sir William Stuart. He made them stare. He sported white shooting-jackets full of pockets and of dandy cut, trousers of shepherd's plaid, a Panama hat! He carried a two-shoot fowling piece slung across one shoulder! The mountain men almost busted themselves laughing at the poor greenhorn's equipment. But worse was to follow: he had wagons — wagons loaded with tins of preserved meat, bottles of pickles, brandy, porter, fine wines, coffee, tea. He dined off hams and tongues, had flour by the barrel, real sugar, and his table service staggered the minds of the simple trappers, whose whole outfit consisted of a skinning-knife and a tin cup. Kit guffawed with the rest at the worthless stuff the man toted.

But before long the trappers changed their tune. Sir William was a splendid horseman; he could shoot with the best of them; and the way he made 'em come in the first Indian skirmish showed what fighters the Scotch could be. Moreover, he was as liberal with his strange food and fancy drinks as the most open-handed of mountaineers. He soon had the trappers eating out of his hand.

It made Kit feel queer as a gut-shot coyote. 'To look at the man, and then to see him ride, and shoot, and fight Injuns the way he done it!' The world was bigger than Kit had thought. For a while he held out, but at last became the firmest of Sir William's trapper friends. That taught Kit a lesson. Men are men, wherever you find 'em. For the first time, Kit Carson opened his eyes to the merits of the greater world on which he had turned his back.

It was time he learned the lesson, for the beaver trade was slumping badly. Beaver was going down. Once a good plew brought six dollars, beaver or kitten. But the value had steadily dropped from season to season, until now a dollar was the standard price at Taos or St. Louis. The mountain man had to find other work to keep his head above water. Traps wouldn't pay for everything he needed now.

The cause of the sudden decline in the price of beaver fur was sufficiently remote from the Rocky Mountains. Up until 1832 all the finest hats had been made of beaver, and it was this demand that made the prosperity of the fur trade. But in that year the silk hat was invented, and immediately it was perceived that the days of the beaver trade were numbered. The silk hat gained vogue slowly, but by 1837 the trappers had felt the slump. Sir William had warned Kit of the coming change, but the mountain men could not believe it.

'Beaver's sure to rise,' said old Bill Williams, sitting in camp on the Uintah. 'It aint in human nature not to trap.' He and Kit and the rest had just arrived at Roubidoux's fort, and had been offered one dollar the plew for their hard-won peltries. If it had been anybody but old Antoine, they would have accused him of cheating and taken their furs elsewhere. But they knew they could trust Antoine.

It was a gloomy group of men who sat round the fire that night. Their hearts were small. Beaver wouldn't pay for trapping now. They looked out upon a world utterly changed for them.

Kit, who faced facts more readily than the rest, was most cast down. He didn't like old Antoine's price at all. It meant that his profession was wiped out. Not yet thirty years old, and his trade gone under. After working and fighting his way to the top of it, to see it go under overnight! He who had been king of the mountains, carousing and dancing and fighting, living on the fat of the land, with camp-keepers and servants to do his dirty work, was now an outcast and a poor man. It was hard. It was incredible. It was true. More deviltry of the cussed greenhorns in the settlements! Hell's full of high silk hats!

This animosity of the mountain man against silk hats and all who wore them persisted long after the end of the beaver trade. Many a tenderfoot in later years had his high hat perforated by bullets from the pistols of indignant Westerners, who had forgotten the historic origin of their dislike. Even to-day your true-bred cowboy will brag that his ten-gallon Stetson is 'genuwine beaver.' The cowboy's hat is a symbol of heroism to millions who have never seen the West: but the silk hat is equally a symbol among the folk for everything that regular guys abhor. This is an American folkway: the Englishman, for example, knows no such feeling for the topper. Unquestionably, the trapper is the father of this popular prejudice.

Well, there sat Kit, Old Bill Williams, Bill New, and Frederick. What were they to do? What *could* they do? Stay in the mountains and starve? Hang on until beaver went up again? Most of them could not conceive of a different life.

But Kit's thoughts immediately turned to the man who had first befriended him — the handsome man in buckskins who had helped him escape from the hated saddler's bench. Charles Bent had held out a hand where one trail ended and another began. Maybe so he would do it again.

'Boys,' said Kit, 'the beaver trade's rubbed out. Let's put out for the big lodge on the Arkansas, and see what Bent and St. Vrain can do for us. Their like were never in the mountains, and if anybody can tell us how the land lays, they can.'

'Wagh!' growled Bill Mitchell. 'If your stick floats thataway, Kit, hyar goes. Catch up, boys; put out. This coon is makin' tracks to the fort come sunup. If Charlie Bent caint help us, thar's no help this side of hell.'

Dawn saw them on the march. Ahead lay the trail which Kit had followed when White Head first led him into the Rocky Mountains. Now he was going back. Back to the south. Back to the same place — to make a new start. On either side rode his comrades — Bill New, the red-headed, Bill Mitchell, actually wearing pants for a change, Frederick, the fat 'Dutchman,' and a swarthy, voluble Frenchman, name unknown. Behind followed Waa-nibe astride a pack-horse, and little Adeline's bright eyes stared out from the hooded babyboard over the horse's tail.

That journey marked an epoch in the history of the West and in the lives of Kit Carson and his companions. Youth and the mountains lay behind. Manhood and the plains lay ahead — fame too, and mastery of a new world. Who were these men who followed Kit on the new trail?

Of Frederick we know nothing. But Ruxton has given us a full-length portrait of Old Bill Williams — which is worth quoting. For of all the mountain men on record, Bill Williams was unquestionably the most notorious and celebrated character.

When Kit and he rode down to Bent's Fort in 1838, Bill was already almost fifty years old, and he had been a long time among the Indians. Bill started life as a Methodist preacher in Missouri. There, according to his own brag, he was so well known on his circuit that whenever he appeared even the chickens in the farmyards would recognize him, and the roosters would begin to crow: 'Hyar comes Parson Williams; one of us goes in the kittle to-day!' Exhorting sinners was very much to Old Bill's taste. But one Sunday morning he saw a likely gal settin' on the front bench at preachin', and durned if she warn't so all-fired fofurraw that the Parson couldn't keep his mind on what he was thinkin' about. So he come down outen the pulpit and said he reckoned he warn't fit to preach the gospel nohow. Immediately, changing his Bible for a rifle, he set out for the Rocky Mountains.

There he moved about from tribe to tribe, an eccentric figure, and there he soon acquired a reputation as a medicine man among the Utes. But in time, as his sacred labors fell farther and farther behind him, he joined a band of trappers. His importance in the Old West is indicated by the monuments that there remain to him: Bill Williams Peak, the Williams Fork of Grand River, Colorado, and the station of Williams, in Arizona.

'Williams always rode ahead, his body bent over his saddle-horn, across which rested a long heavy rifle, his keen gray eyes peering from under the slouched brim of a flexible felt hat, black and shining with grease. His buckskin hunting-shirt, bedaubed until it had the appearance of polished leather, hung in folds over his bony carcass; his nether extremities being clothed in pantaloons of the same material (with scattered fringes down the outside of the leg — which ornaments, however, had been pretty well thinned to supply "whangs" for mending moccasins or pack-saddles), which, shrunk with wet, clung tightly to his long, spare, sinewy legs. His feet were thrust into a pair of Mexican stirrups made of wood, and as big as coal-scuttles; and iron spurs of incredible proportions, with tinkling drops attached to the rowels, were fastened to his heel — a bead-worked strap, four inches broad, securing them over the instep. In the shoulder-belt, which sustained his powder-horn and bullet-pouch, were fastened the various instruments of one pursuing his mode of life. An awl, with deer-horn handle, and the point defended by a case of cherrywood carved by his own hand, hung at the back of the belt, side by side with a worm for cleaning the rifle; and under this was a squat and quaint-looking bullet-mould, the handles guarded by strips of buckskin to save his fingers from burning when running balls, having for its companion a little bottle made from the point of an antelope's horn, scraped transparent, which contained the "medicine" used in baiting the traps. The old "coon's" face was sharp and thin, a long nose and chin hob-nobbing each other; and his head was always bent forward, giving him the appearance of being hump-backed.

He *appeared* to look neither to the right nor left, but in fact his little twinkling eye was everywhere. He looked at no one he was addressing, always seeming to be thinking about something else than the subject of his discourse, speaking in a whining, thin, cracked voice, and in a tone that left the hearer in doubt whether he was laughing or crying. . . . His character was well known. Acquainted with every inch of the Far West, and with all the Indian tribes who inhabited it, he never failed to outwit his red enemies, and generally made his appearance at the rendezvous . . . with galore of beaver, when numerous bands of trappers dropped in on foot, having been despoiled of their packs and animals by the very Indians through the midst of whom old Williams had contrived to pass unseen and unmolested. On occasions when he had been in company with others, and attacked by Indians, Bill invariably fought manfully, and with all the coolness that perfect indifference to death or danger could give, but always "on his own hook." His rifle cracked away merrily, and never spoke in vain; and in a charge — if it ever came to that — his keen-edged butcher-knife tickled the fleece of many a Blackfoot. But, at the same time, if he saw that discretion was the better part of valor, and affairs wore so cloudy an aspect as to render retreat advisable, he would first express his opinion in curt terms, and decisively, and charging up his rifle, would take himself off and *cache* (hide) so effectually that to search for him was utterly useless. Thus, when with a large party of trappers, when anything occurred which gave him a hint that trouble was coming, or more Indians were about than he considered good for his animals, Bill was wont to exclaim —

"'Do 'ee hyar now, boys, thar's sign about? This hoss feels like caching"; and without more words, and stoically deaf to all remonstrances, he would forthwith proceed to pack his animals, talking the while to an old crop-eared raw-boned Nez-Percé pony, his own particular saddle-horse, who in dogged temper and iron hardiness, was a worthy companion of his self-willed master. This beast, as Bill seized his apishamore to lay upon its galled back, would express displeasure by humping its back and shaking its withers with a wincing motion, that always excited the ire of the old trapper; and no sooner had he laid the apishamore smoothly on the chafed skin, than a wriggle of the animal shook it off.

"'Do 'ee hyar now, you darned critter?" he would whine out, "can't 'ee keep quiet your old fleece now? Isn't this old coon putting out to save 'ee from the darned Injuns now, do 'ee hyar?" And then, continuing his work and taking no notice of his comrades, who stood by bantering the eccentric old trapper, he would soliloquize—"Do'ee hyar now? This nigger sees sign ahead — he does! he'll be afoot afore long, if he don't keep his eyes skinned — *he* will. *Injuns* is all about, they are: Blackfoot at that. Can't come round this child — they can't, wagh!" And at last, his pack-animals securely tied to the tail of his horse, he would mount, and throwing the rifle across the horn of his saddle, and without noticing his companions, would drive the jingling spurs into his horse's gaunt sides, and muttering, "Can't come round this child — they can't," would ride away; and nothing more would be seen or heard of him perhaps for months, when they would not unfrequently, themselves bereft of animals in the scrape he had foreseen, find him located in some solitary valley, in his lonely camp, with his animals securely picketed around, and his peltries safe. . . .

'His iron frame defied fatigue, and at night, his love for himself and his own animals was sufficient guarantee that the camp would be well guarded. As he rode ahead, his spurs jingling and thumping the sides of his old horse at every step, he managed, with admirable dexterity, to take advantage of the best line of country to follow — avoiding the gullies and canyons and broken ground, which would otherwise have impeded his advance. This tact appeared instinctive. In selecting a camping-site he displayed equal skill; wood, water, and grass began to fill his thoughts towards sundown; and when these three requisites for a camping-ground presented themselves, old Bill sprang from his saddle, unpacked his animals in a twinkling and hobbled them, struck fire and ignited a few chips (leaving the rest to pack in the wood), lit his pipe, and enjoyed himself.'[1]

On the Arkansas, Mitchell and New dropped out, wishing to postpone the inevitable surrender as long as possible, saying that the hunting looked good thereabouts. Kit, Williams, Frederick, the Frenchman, and Waa-nibe rode on downstream a hundred miles and arrived in time within sight of the gray adobe block — Bent's Old Fort.

Before the great gloomy *portal* they halted. Then, riding in, Kit got down from his buffalo horse, and, standing in the sunlight of the white-washed *patio*,

---

[1] Ruxton, *In the Old West*, pp. 184–89. Quoted by courtesy of The Macmillan Company. The original edition in book form bore the title, *Life in the Far West*.

asked for Charles Bent. Soon after the trader came through the dark doorway of his storeroom, his moccasins scraping the gravel, and grasped Kit by the hand. Charles Bent did not disappoint the trappers. He engaged Kit as hunter for the fort — at wages of one dollar a day.

Kit pitched his tepee alongside the fort. Waa-nibe could remain there for the time being. Kit began to make plans. He would have to ride down to Taos and recruit men for the work ahead.

But before he left, two recruits came in, eager for work. Mitchell and New walked down-river a hundred miles, stark naked. The Indians had found them in camp on the Arkansas, had taken everything they possessed except their hair, and left them to walk home barefoot over the plains. New was red all over when he got in: red-headed, red from the sun, red from anger, red from shame at the laughter of the squaws who watched from the roofs about the graveled *patio!*

Bent's Fort was a large establishment, and in winter the staff comprised a large number of traders, clerks, hunters, packers, teamsters, laborers skilled and unskilled, and servants, of Spanish, Mexican, French, German, Indian, negro, and American blood. Somewhat more than a hundred men were employed there, and most of them had wives and children. And Indian wives always had relatives with wives and children. And there were always guests: trappers in to sell a few peltries, men from the settlements, sportsmen, invalids, soldiers, explorers.

All those people had to be fed — and fed on meat. And the meat had to be found, killed, butchered, jerked, hauled to the fort. If by any chance it could be killed near by, it might be brought in fresh and stored in the big ice-house on the river-bank: for, of course, Bent, being a Yankee, had an ice-house — a luxury without a rival on the plains.

Now all these people led active outdoor lives and had appetites like ostriches. Moreover, the staple food was buffalo meat, and it was a proverb on the prairies that a man could eat his own weight of buffalo meat and never have a bellyache. Instances are on record of Indians — professional feasters — who made away with forty pounds of it in one day. Allowing for exaggeration, it is clear that nobody at Bent's Old Fort would be satisfied with what a modern hotel menu dignifies by the name of a portion of beefsteak. It took a good many bison to fill those hundreds of mouths under Bent's hospitable roof.

Buffalo bulls were too tough to eat. The cows only were killed — except in time of famine. Now, a buffalo bull weighs rather less than a steer on the hoof, and the cow less than the bull. Dressed, the buffalo cow will weigh a good deal less than the steer, even when handled by a butcher in a packing-house. On the prairie, where the waste was greater, the disparity was marked.

During the summer months most of Bent's people were away with the wagon-trains going to Missouri and returning. But for seven months in the year Kit and his men had between two and three hundred mouths to fill, not counting the Indians encamped about the fort who partook of its hospitality. To feed all these people, Kit adopted Indian methods of hunting.

The Plains Indians in old times followed the primitive method of driving or luring the bison over cliffs or into traps. But with the advent of the horse such methods fell into disuse. Approaching, still hunting, or stalking was another method — employed when a single hunter, or a small group, needed a small supply of meat.

In approaching, the hunter rode as near the herd of cows as possible without alarming them — facing the wind all the way, of course, for although the buffalo had poor sight and hearing, its nose was very keen. Dismounting and hobbling his horse (taking care to leave the lariat trailing so that he could easily catch the animal again), the hunter crept forward within range, taking advantage of any cover there might be, and fired into the herd until he had bagged his game, or until the animals took fright and ran away.

Unless shot through lungs or heart or spine, the buffalo took a tremendous amount of killing. It was generally careless of danger, and a man could fire repeatedly into a small herd without dispersing them. Only in the rutting season was there any great danger from angry bulls. Indeed, the bulls became so lethargic after drinking that on one occasion Kit actually walked up and laid his hand on their rumps.

In firing, the hunter aimed at a point behind the shoulder and just a few inches above the brisket — the only vulnerable point. Otherwise, he might flatten his bullets against the buffalo's broad forehead all day without effect, or shoot at the huge carcass forever with no more result than the loss of his lead.

But approaching — even with the best of luck — would never have accounted for the number of animals required to feed Bent's Fort. It was useful on occasion — about the fort, or with the wagon-train.

But the third method — called running meat — was the one practicable way of killing large numbers of bison at one time. Kit adopted this method — the Indian method.

The Plains Indians made two big hunts each year. In September, when the cows were fat and the wool was fresh and thick, they set out from their villages with small traveling lodges and all the active members of the tribe and kept going until they found the herds. Then, having killed enough meat and obtained enough robes for six moons or so, they turned back and rode home again. Again, in April, the same long journey was taken, and the same hunt was made, except that this time the robes were worthless, and only the hides were taken. For in summer the buffalo was shedding. These two hunts occupied about half the year — sometimes more — for the Indians might travel as far as three hundred to a thousand miles on their way to and from the herds.

The buffalo migrated up and down the plains with fair regularity spring and fall, but many things — a fire, a flood, a stampede, a hunt — might turn them aside, and nobody could tell with certainty just where to find them. The buffalo could not be depended upon to come to the door of the tepee. The Indians had to go and find them.

They set out in the likeliest direction, throwing scouts far ahead, scouts whose business it was to find the herds and report back to the main camp. When this happened, the main camp was moved as near the herds as possible — say to a distance of twenty miles or so; for great care was taken not to frighten the buffalo away. The camp was policed by societies of warriors called 'soldiers,' and these 'soldiers' patrolled the country to see that no man slipped out of camp to hunt alone. Such a hunter might scare away the herds and cause the whole tribe to suffer privation. When the 'soldiers' caught such a man, they lashed him thoroughly with their quirts, cut his robes and lodge to pieces, killed his horses, and raised hell with him generally for his selfishness.

As soon as the camp reached the hunting grounds, the Indians — every available man — rode out to make the surround. Each man led his trained buffalo chaser, riding some less valuable horse, or running on foot. For success in running meat depended upon the speed and agility of the horse, and every Indian kept an animal especially trained for this work. An ordinary horse could not be induced to go near a buffalo.

On the way to the herds, the 'soldiers' surrounded the other men, keeping a keen eye out for any who might slip off ahead of the main body. And when the herd was reached, they halted the men, and they all mounted their buffalo ponies.

The party then spread out to get on all sides of the herd, and at a signal, all charged into the herd from different directions. This kept the buffalo milling round and round until the mounted hunters had killed all the animals — or as many as they could use. By that time the women had come up with pack-animals and travois, and the pack-trains went home loaded with rich, red, juicy beef.

Riding into the herd was a somewhat dangerous proceeding. The cows were swifter than the bulls, and so the hunter had to force his way through the bulls to reach his game. The dust was smothering, and there might be prairie-dog or badger holes into which a horse might step, breaking his leg and throwing his rider under the hooves of the stampeding bulls. To fire effectively, the hunter had to ride alongside. And with the old-fashioned muzzle-loader, there was every danger — after the first shot — that the ball (never rammed down) might roll toward the muzzle just as the gun was fired. In that case, an explosion followed that might smash gun and man. Broken Hand got his name from just such an accident.

The work of butchering required much skill. Often the butchers worked far into the night to save the meat from the wolves which loped hungrily about. As soon as a fat cow was killed — for the bulls were too tough to eat — the hunter dismounted, tied his horse to the horns of the game, and if possible turned the cow upon her belly. To maintain the carcass in that position, he stretched out the legs to the side, and sometimes pulled the head round to prop the shoulder.

Then he got out his knife and made a transverse slit across the nape of the neck. Taking hold of the hairy lump known as the boss, he quickly sliced it off with the hide attached. The boss was the size of a man's head. It projected from the back of the neck just behind the shoulders, and when boiled tasted very much like marrow — very tender, very nutritious, very appetizing.

Another cut was made along the spine from the boss to the tail, and the skin was separated from the body on each side, pulled down to the brisket (to which it was still attached), and stretched flat upon the ground, ready to receive the meat. By laying the meat on the skin, the hunter kept it clean. Then the butchering began.

First the shoulder was severed, and the fleece removed from along the backbone. The fleece was the thin layer of flesh covering the ribs. That on the belly was known as the belly-fleece.

After that the hump-ribs — those vertical projections of the vertebræ above the spine — were chopped off with a hatchet, or broken off with a leg-bone severed at the hock and used as a mallet, in case no hatchet was at hand. It may seem very unobservant of the trappers to call those projections ribs. But no one who has seen a buffalo butchered will think it strange. The hump-ribs were flat, and often more than a foot in length.

The back-fat, or *depuis*, the broad fat part extending from hump to tail, was next removed and laid upon the skin. Afterward the tenderloin and tallow were secured by opening the animal, and generally the *boudins* (the intestines containing the chyme) were allowed to tumble out. *Boudins*, emptied, reversed, washed, tied at the ends, and broiled over a bed of coals, made a most delicious dish, as the rich fat was all cooked inside. It was a favorite treat of the trappers.

From the head the tongue was cut out. Then the hunter was usually content, leaving the shoulders, hams, side-ribs, and head to be cleaned up by the wolves and coyotes. However, in time of famine every portion was saved, the bones and even the hoofs boiled and eaten. But that was considered hard doin's by the mountain men.

Among the Indians, two men generally worked together on the same animal, and the speed with which they could reduce a carcass to red ruin has often been commented upon. Twenty minutes to half an hour sufficed, though, if hungry, the hunters might pause in their work to share a portion of warm, raw liver seasoned with gall. All who came up were permitted to help butcher, and each man received a certain part for his labor — a part arbitrarily determined by the order in which the men went to work. The hide and tongue usually belonged to the man who killed the cow.

Having stripped the skeleton of its meat, the hunters next had to pack it in the skin on muleback and go back to camp — a distance of twenty miles or so — with the load. There the meat had to be jerked at once, or all of it would spoil within a few days.

Jerking or making meat was a very important process and could not be postponed. The meat was cut into thin sheets and hung out to dry like so much washing. Some of these sheets might be no bigger than your hand; others — parts of the fleece, for instance — might be as large as a face-towel, and stretched upon long skewers about the size of lead-pencils. All these sheets were hung upon lines and scaffolds out of reach of wolves or dogs, and left to dry in sun and wind. During this process — and in butchering — the meat was *never* cut across the grain.

In the high, dry air of the plains making meat did not take long — if the weather was favorable. But if it rained or was damp, the meat had to be dried over a fire on a scaffold of sticks built like a gridiron and raised several feet above the ground. In damp weather there was always danger that the meat might become fly-blown and spoil, and then all the labor of the hunter would be lost.

Greenhorns were inclined to slight this process, and as a result hundreds of them died of dysentery. The trappers always had time to make meat properly, however pressing their haste.

When dry, the sheets of meat — stiff as a board — could be packed on muleback or stacked in a wagon like so many shingles, and transported readily enough. However, if the dried meat was to keep long even then, it had to be pulverized with a stone hammer (the labor of women) and mixed with powdered cherries and tallow tightly packed in a rawhide case. It was then called pemmican, and would keep indefinitely, if dry.

From this account of the work of the buffalo hunter, it will be perfectly clear that running meat was no pastime for an idle afternoon. First there was the tedious business of locating the herds; then the long march to their vicinity, the making camp, the cautious advance to the surround, the dangerous business of riding into the herd and shooting the fat cows — cows which could run as fast as a horse. Then there was the butchering, hot work in the open, work that could not wait. And then there was the packing of the meat on muleback, the long march back to camp, the slicing of the meat that night and hanging it up to dry — perhaps the keeping up of fires under it for four or five days. And lastly, the long ride home to the fort with the jerked meat.

Some biographers speak lightly of Kit working as hunter for the fort, as though he could kill a buffalo on the doorstep and carry it in on his shoulder. The job of hunting for Bent's Old Fort was a business that called for the labor of all the Carson Men twice a year — a business that kept half a hundred men busy for weeks at a time, for, on the average, a buffalo cow would make only one hundred pounds of dried meat.

That autumn of 1838 Kit made his first big hunt for Bent. Who were the Carson Men who helped him run meat?

# III
PLAINSMAN

CHAPTER XIV
THE CARSON MEN

KIT soon found that he could have the pick of the mountain men for his band of hunters. They all respected his integrity and competence, and now that beaver trapping had failed, the trappers clung together, distrusting the men of other walks in life. Moreover, Kit's was the only year-round employment to be had in the Old Southwest at that time. He maintained his organization of mountaineers for about fifteen years — certainly until the blowing up of Bent's Old Fort in 1852. During those years there must have been many changes in the personnel. For the mountain man was a rover, and as times changed and the tide of travel began to sweep across the continent, Kit's men were in constant demand as guides and scouts. Yet, even then, most of them would prefer Kit as their *bourgeois*, for trapping and hunting were their chosen jobs. It is probable that almost every qualified mountain man in the Southwest was at one time or another a member of Kit's band.

They were all seasoned trappers — Americans from Missouri, Kentucky, Ohio, Tennessee — with a few French Creoles, a sprinkling of Spaniards, several Delaware Indians. They were all men who shot straight, thought straight, talked with a straight tongue. They prided themselves upon the fact that they were fightin' men — that their enemies might kill them, but could never make them give in. They made that boast good.

In spring and fall Kit led them on the buffalo hunt for Bent's Old Fort. And from that fort, or from Taos, where he later made his home, Kit sent out those trapping brigades, those bands of riders to guard the Santa Fe Trail, those punitive expeditions against the Indians, those rescue parties that have made the Carson Band famous throughout the West.

One incident — hitherto unpublished — will illustrate the character of these men, and the temper of the man who led them.

In the summer of 1839, Kit and his band were patrolling the Santa Fe Trail, guarding the route against the marauding Pawnees, who made life a burden to the wagon-masters. Bent and St. Vrain were anxious to keep the trails open, as their business could not go on otherwise, and they often sent Kit up and down the route — even as far east as Pawnee Rock. On this occasion, the Carson Band was encamped on the Arkansas, not far from Chouteau's Island.

Kit had gone out with two companions to shoot antelope, and had left Solomon Silver in charge of the camp. Early in the afternoon, Silver saw a dust to the south of the river. It approached, and a party of Indians appeared. They crossed the shallow river, riding directly toward the camp. Silver saw they were Kiowas, who traded with Bent and St. Vrain, and therefore allowed them to come in. It was a large party — some fifty warriors — led by the chief Dohau-sen, sometimes called Sierrito, or Little Mountain.

Silver invited the chief to sit and smoke with him.

Do-hau-sen assented, and Silver took a plug of tobacco from his possible sack, cut up a sufficient quantity, and packing it in the bowl of the pipe, struck steel on flint and soon had the smoke going. While the pipe passed, he asked where Do-hau-sen was bound.

The chief explained that he and his party were out for Cheyenne hair. Now Silver knew that Bent (and of course Kit) were always trying to make peace between the Cheyennes and the Kiowas. For both tribes traded with Bent, and the enmity between them was a source of constant trouble and expense to the traders, who had to maintain forts in the range of each tribe. Silver therefore proposed to the Kiowas that it would be better for everybody, if Do-hau-sen would lead his war party against the Pawnees, and leave the Cheyennes alone.

This proposal seemed to the Kiowas a piece of unnecessary effrontery. They had not asked Silver's advice, and they did not think their plans were his to meddle with. Glancing around the small circle of white men, Do-hau-sen grinned. The old chief had a sense of humor. He decided to put the Carson Men in their places. He spoke to Silver.

'My friend,' he said, 'I will go against the Pawnees if your men can count more *coups* than mine.' The old chief thought that deeds were the only warrant for words — an opinion which sadly needs enforcing in our times.

Sol Silver had only twenty-three men in camp, and of these two or three — packers and cook — would have no *coups* to count. For a moment Sol was stumped. The old chief grinned. But Sol was not the man to take a dare. He assented, and the circle was enlarged to allow the white men to sit in. Then the contest began.

Counting or matching *coups* in this way was a common amusement of the warriors, and every man knew exactly what his fellows were entitled to claim in the way of war honors. Sol and the trappers could judge pretty well from the way in which the warriors were painted and feathered as to whether or not they were telling the truth. And Do-hau-sen could easily count the tacks or notches on the stocks of the white men's rifles.

Anticipating an easy triumph, Do-hau-sen declined to lead off, and told his youngest warriors to begin the game. One by one they stood up, and with graphic gestures narrated their deeds of valor — against the Mexicans, against the Pawnees, against the Cheyennes, the Arapahoes, the Osages, the Shawnees and Delawares, against the whites. Satank was there, with his lean Mongol jaw and scrawny mustache; Satanta, the high-spirited orator; Stumbling Bear, then a young man, but with the broad brow, heavy jowl, and fierce eyes which appear in his later portraits; Yellow Hair, Eagle Feather, Hunting Horse; one after another they got up, told their stories confidently, and sat down. They were all certain of a smashing victory over the presumptuous whites. Do-hau-sen said he did not think he would need to rehearse *his* brave deeds: let the white men talk.

Then Sol called upon his trappers.

First there was Oliver Wiggins, Kit's latest recruit, long and lank, going on seventeen, who told, with boyish modesty and boyish pride, how he had stood off the Comanches last summer. Oliver Wiggins, the very name a parable of the homely heroism of the Old West.

Honus, LeFevre, Pilka, Tim Goodale; Canty, the drawling youngster from Ohio; Rube Herring, towering even above the haughty Kiowas. Dick Wooton, Kit's favorite giant, the young Lochinvar of the vale of Taos, told boldly how he made 'em come.

Then the Delawares stood up: Tom Hill, with hair that hung below his knees; Black Beaver, best shot in all his tribe; Jonas. After them, Chabonard, the breed, son of Sacajawea, graceful, urbane, fluent; Charlie Otterby, getting on, but still French and debonair; Bill New, with the fiery red head; Bill Mitchell, Hawkins, Markhead, spoke and sat again. Tom Tobin; and Ike Chamberlin, heavy and red of face, whom the Indians knew as Pä-bote, American Horse, because he was too large to ride a pony and had to import a saddler from the States. Big, intolerant, gallant boys, they wished they could count a thousand *coups* that day: the credit of Kit's band was at stake.

Last of all, Old Bill Williams sawed the air, parson-like, and squeaked out his tally of good Injuns. Then he sat down, cross-legged, grumbling and sucking his old black clay. The cussed Injuns couldn't come round that child: he knowed how to fetch 'em, and done it, too.

At the end the Kiowas were still ahead, and Do-hausen sardonically signaled Silver, whom he knew as Tonguadal, a name whose English translation (and implication) is unhappily quite unprintable. Silver got up. He knew every Kiowa there, and they had good reason to remember him. He did not trouble to remove the buckskin cover from his rifle. But as the Indians looked up at the swarthy, muscular, Spanish man, with his black hair and beard bushed out like a pirate's, they did not doubt that he would have a story to tell. At that, Sol surprised them.

Sol, who never knew his real name, was carried off from some town in Old Mexico (Saltillo, he believed) by the raiding Comanches, who murdered his parents, and then sold the boy to the Kiowas. The Kiowas were always on the lookout for captive children: they were few in number, and had to recruit where they could. Sol used to brag that his Kiowa owner paid for him four butcher-knives, two buffalo robes, and a plug of tobacco. Sol did not like his new master, who sent the boy out on the cold hills to herd his ponies. Sol almost froze there, not having any clothing but a scrap of old lodge-cloth. A Kiowa boy — an orphan — shared his vigil, and kept an eye on Sol. Finding that he could outwrestle the Kiowa boy, Sol got possession of a knife and killed him. Then he jumped on his master's best horse and rode away. He rode hard, thinking the Kiowas were hot on his heels, and as a result, his pony gave out on the Arkansas. There he was seen by a wagon-train, and the men would have shot him. But the wagon-master, seeing it was only a boy, protected him and carried him to Missouri. Sol soon left the settlements and threw in with the Osages near by, and — according to his own account — became a famous warrior. When they organized a party to go against the Kiowas, Sol gladly went along.

The Osages struck the Kiowa camp on Otter Creek near the Wichita Mountains in what is now Oklahoma, and surprised it when many of the warriors were gone.

The Osages destroyed the camp and nearly every soul in it, and all without the loss of a man. Naturally, they were delighted. In the tepees they found a lot of brass kettles, and in their high spirits decided that it would be a good joke to place these kettles in a ring around the burned camp, with a Kiowa head in each kettle. In this fight, Sol counted *coup* upon, killed, and scalped his old master. This happened about the time of the great meteor shower (1833) — the year it rained fire.[1]

Among the Osages, Sol became so successful a warrior that they tattooed the mark of honor on his chest. This mark — a design of converging blue diagonals — was a cause of great pride to Silver. In a fight with Indians, he was accustomed to strip off his shirt, so that the redskins might see the mark and lose heart. In his ears he wore the earrings of his Kiowa master — large rings of *silver*, from which he took his name.

Sol began to narrate his deeds of bravery in battle, but he saw that the Kiowas were so far ahead that his tally would not add enough to beat them. He had not mentioned the affair of the brass kettles, not knowing how the Kiowas would take it. But now, being on his mettle, he took the risk, stripped off his shirt, displayed the mark of honor, and told them how he had charged into their camp that day. Pointing to his earrings, he reminded them that these had once belonged to one of their tribe.

'Look,' he said, 'do you savvy these? *I* touched him; *I* took his hair.'

Sol watched the chief to see how he would take that.

[1] A full account of this fight will be found in Mooney, *A Kiowa Calendar*, Seventeenth Annual Report of the Bureau of American Ethnology.

Do-hau-sen sat unmoved. He may have liked the story, for it was that disaster which had led to the deposition of his predecessor and his own elevation to the chieftainship. The tally was still short, and Sol Silver, knowing that the Kiowas could not check his statements, kept adding Kiowa after Kiowa to his record, allowing his imagination free rein, until he had made an imposing total. The Indians began to talk among themselves.

Do-hau-sen got to his feet. But before he could speak, Kit and the antelope hunters rode in. At once, Kit asked what was going on, and being told the state of the case, proposed that he also be allowed to sit in. Having smoked, he sat waiting for Do-hau-sen to speak. The old chief began, recited a lengthy list of honors. And as often as he told how he struck the enemy, the Kiowas yelled *Ho, Ho*, and beat upon the ground with their weapons. Then he sat down.

Kit was the last. He told them of the massacre of the Apaches under Ewing Young, *Coup* No. 1; of the village of hostiles he destroyed in California for the *padres* of the Mission, *Coup* No. 2; of the eight horse-thieves he helped to rub out, *Coup* No. 3; of the Crow whom he killed near Bent's Old Fort, *Coup* No. 4.

There was the Indian he dropped while besieged in Gaunt's log house on the Arkansas, and another in the Bayou, who came to stampede Gaunt's horses, *Coups* Nos. 5 and 6.

There was the Indian who stole Robidoux's horses, *Coup* No. 7. There was the Blackfoot whom he shot to save Markhead, the time he himself was so severely wounded, *Coup* No. 8.

There was the battle with Comanches south of the

Arkansas, *Coups* Nos. 9 and 10. There was the best *coup* of all — the duel with Shunar — No. 11. There was 'the prettiest fight,' when Kit saved Cotton and Doc Newell and came near losing his own hair, *Coups* Nos. 12, 13, 14. There was the fight against Blackfoot fire, *Coup* No. 15. There was the battle with the Blackfoot on the island in the Yellowstone, *Coups* Nos. 16 and 17.

And finally, there was a little fracas the previous summer, when Kit had defended a wagon-train against the Kiowas themselves. He showed them the brass tacks in his rifle stock. Eighteen of them.

Then Kit sat down. That was all he had to show for his ten years in the mountains.

Do-hau-sen was beaten, and without a word his crestfallen braves mounted their ponies and rode away. Do-hau-sen shook Kit's hand in parting, and spoke to him in the language of signs.

'Little Chief, you are my friend, and I am yours. *Muy amigo*. You have won. You and your men have counted more *coups* than all my warriors. But I am going to lead my warriors against the Cheyennes all the same. While you were out hunting antelopes, my friend Ton-guadal here had already killed more Kiowas than there were in the camp on Otter Creek that day. Ton-guadal is big medicine: he kills Kiowas just by talking. He is a great man, a great warrior — and a great liar! I think I had better go now, or he will kill me too!'

With that Do-hau-sen mounted his pony, and, grinning back at Sol, rode off. It was a long time before Sol Silver heard the last of that.

Such were the Carson Men, so far as their names are

known. Their deeds are half forgotten. But they were not the least of the builders of the nation. It was no small honor to Kit that he was able to command such fighters. ... And what was Kit Carson like at this time of life? Let Ruxton give a contemporary portrait:

'Last in height, but first in every quality which constitutes excellence in a mountaineer, whether of indomitable courage or perfect indifference to death or danger — with an iron frame capable of withstanding hunger, thirst, heat, cold, fatigue, and hardships of every kind — of wonderful presence of mind and endless resource in time of peril — with the instinct of an animal and the moral courage of a *man* — who was "taller" for his inches than Kit Carson, paragon of mountaineers? Small in stature and slenderly limbed, but with muscles of wire, with a fair complexion and quiet intelligent features, to look at Kit none would suppose that the mild-looking being before him was an incarnate devil in an Indian fight, and had raised more hair from head of Redskins than any two men in the western country; and yet, thirty winters had scarcely planted a line or furrow on his clean-shaven face. No name, however, was better known in the mountains — from Yellowstone to Spanish Peaks, from Missouri to Columbia River — than that of Kit Carson, raised in Boonlick, Missouri State, and a credit to the "diggins" that gave him birth.' [1]

Wagh!!

---

[1] Ruxton, *In the Old West*, pp. 286–87. Quoted by courtesy of The Macmillan Company. The original edition in book form bore the title, *Life in the Far West*.

CHAPTER XV
THE CHEYENNE WOMAN

ABOUT the time of Kit's marriage to the Arapaho girl, William Bent also trapped a squaw. This was Owl Woman, a young Cheyenne. She was the eldest daughter of Gray Thunder, keeper of the Cheyenne Medicine Arrows, and therefore the most influential man in the tribe.

The medicine arrows are the most sacred and prized talismans of the tribe. Their origin goes back to the mythical culture hero, and they were in the possession of the Cheyennes when white men first met the tribe. Even to-day, when missions and Government schools have done their worst, no Cheyenne will so much as name these sacred arrows, and no white man has ever been permitted to inspect them. Elaborate rituals kept them from the touch of the profane, and it was believed that so long as they were treated with reverence, the Cheyennes would remain happy and powerful.

Of course, the keeper enjoyed great honor, and on most occasions his word was sufficient law unto his people. Bent was therefore allying himself with the Cheyenne royal family — or at least the family of the high priest. Bent must have been strongly attracted to this girl, or must have anticipated great advantages from the alliance. For as a rule traders preferred to marry women of tribes at war with their customers, in order to avoid the heavy burden of entertaining the relatives of their wives.

Bent constantly impressed upon his father-in-law by talk and gifts and influence of every kind the extreme desirability of peace between the Cheyennes and other tribes with whom he traded. Apparently, the old man was swayed to favor the views of his white son-in-law. But the Cheyennes were always a haughty, headstrong people, and even Gray Thunder found at times that he could not control the young men.

Whenever a serious crime — such as murder — was committed within the tribe, it was necessary to renew the medicine arrows in order to placate Man Above and other deities. Renewing the arrows was a tedious ceremony lasting four days. It meant attaching new feathers to the four shafts, and re-attaching the four stone points with fresh sinews. Only the keeper could authorize such a renewing of the sacred arrows.

Until the arrows had been renewed, it was believed that the curse of the crime hung over the tribe. No one dared undertake anything of importance, fearing bad luck. Soon after Bent's marriage a murder was committed within the tribe, and the arrows had to be renewed again.

'It happened at this time that the Bow String soldiers were anxious to go to war. They wished the arrows to be renewed so that they might set out at once, but when they spoke to Gray Thunder, the arrow keeper, about it he told them that the time and place were not propitious and advised them not to go. There was much dispute about this, but at length the Bow String soldiers told Gray Thunder that he must renew the arrows. He refused; whereupon, the soldiers attacked and beat him with their quirts and quirt-handles until he promised to renew the arrows for them. Gray Thunder was then an old man, over seventy. He renewed the arrows as ordered, but before the ceremony he warned the Bow String men that the first time they went to war they would have bad fortune.'[1]

It happened as he had prophesied. The entire party of Bow String warriors was rubbed out on the Washita River by the Kiowas, Comanches, and Apaches. This terrible disaster called for retaliation, and the Cheyennes moved in a body against the enemy. Gray Thunder was leader, but before he could perform the ceremonies to insure a victory, some impatient young men made a premature attack and spoiled his plans. There was a great battle on Wolf Creek in which most of the best men on both sides lost their lives. During the fight, Gray Thunder, stung to fury by the disrespect shown the sacred arrows and the indignities heaped upon him, deliberately threw his life away. A Cheyenne himself, he could be as fiercely rebellious as any of them. 'He said, "I will now give the people a chance to get a smarter man to guide them. They have been calling me a fool."'[2] The old man recklessly exposed himself, and the Comanches and Kiowas rode him down. This was in 1838.

All Bent's plans were knocked into a cocked hat by this disaster. For some time he had been maintaining subsidiary trading forts in the Kiowa and Comanche country at great trouble and expense, simply because those Indians dared not come to trade at Bent's Old Fort on the Arkansas, where they would be sure to find Cheyennes and Arapahoes encamped. And now — just when it had looked as if he might get them to make

[1] Grinnell, *The Fighting Cheyennes*, p. 42.   [2] *Ibid.*, p. 56.

peace — they had fought the biggest battle of all, had killed his father-in-law, had crippled his influence with the tribe.

While William Bent was mourning for Gray Thunder and the smashing of his plans, Kit Carson came back to the fort from the fall hunt and found Waa-nibe dying. A fever had wasted her terribly. The medicine men had tried herbs and incantations. They had sat at her side, holding her pulse and beating the tom-tom in exact time to the pulse, trying, by imperceptibly slowing the drumbeat, to bring back the beat of the pulse to normal. Kit drove them all out and sent for William Bent, who knew more of medical matters than any one in the Indian country. Bent came, and tried, and shook his head. He went back to the fort, and left Kit with his young wife and little Adeline.

Waa-nibe knew she was dying as she lay there on the neat bed of buffalo robes with the trim mats of peeled, painted willow rods hanging from their tripods at head and foot. It was her marriage-bed; she had made it herself, peeling the green willow shoots with her teeth, straightening them in the same way, so that small marks indented their smooth surfaces everywhere. She loved the stocky, inarticulate little man with the blue eyes and sandy hair. He had fought for her, had been the father of her baby, had always been kind to her. Now that her man was back, she feared no evil spirits of sickness. It was not her gods nor his she wanted, but his strong, firm hands about her, and the strange happiness of kissing that he had taught her. Even when he was far off running buffalo, her lips were still on his lips, she told him. . . .

That night, among the dim tepees beside the careless, drifting sands of the Arkansas, the quiet of the moonlit prairies was suddenly broken by a long wail, taken up, repeated, prolonged again and again, and Waa-nibe's aunts and sister rent their clothing, cut off their hair, gashed their limbs with knives until the blood flowed into their moccasins. They stared at Kit, when he came out of the lodge with little Adeline, who cried sympathetically, not knowing what it was all about. Stared at him, asked why he did not wail too. And then Kit had to explain that he was crying in his heart, after the manner of white men.

In the morning they carried the long bundle of buffalo robes down-river, placed it upon a platform in a cottonwood, which rustled soothingly, though no breeze stirred. There they burned the lodge, the bed, all her pretty possessions, and there her brother shot a dog and Waa-nibe's favorite saddle-horse for her journey over the trail of ghosts. And Kit went back to the fort, wondering what he was to do with the baby. He had lost his woman and his heart was sore.

The negress, Charlotte, who boasted that she was 'de onlee lady in de dam Injun country,' tried to cheer him with States doin's — flapjacks, punkin pies, coffee, light bread. And Chipita, the housekeeper, took the baby under her ample wing, uttering soft Spanish syllables and poking its little ribs with a fat brown finger. All the women in the fort sympathized. And Kit left the child to their care after a time and led his men back to Taos.

The fortunes of William Bent and Kit Carson marched together. As both had cause to mourn in 1838, so both rejoiced in 1840. As it happened, the drawn battle on Wolf Creek brought the Indians on both sides to their senses, and after a time they made offers of peace to each other. In the end they all gathered along the Arkansas in the bottoms below Bent's Fort — the largest Indian camp on record — and a lasting peace was made — a peace that has never been broken.

The story of that peace treaty and of all the events that led up to it has been told most graphically:[1] how the Cheyennes and Arapahoes almost bought out Bent's stock in order to make presents to their old enemies; how the Comanches, Kiowas, and Apaches begged the others *not* to give them horses, since already they had more than they could wrangle; how the Cheyennes fired off their guns before presenting them to their allies, so that it sounded as if a battle were in progress; how the Comanches invited the Cheyennes and Arapahoes to visit them afoot and sent every person in both tribes home horseback and driving five or six head each.

Bent's traders were kept very busy during the weeks of that great camp, and Kit's men were charged with the duty of seeing that no one sold any liquor to the Indians. Smuggling was a game much indulged in by the Mexicans along the Arkansas River, but on this occasion Bent was determined that the peace should stand, and that no whiskey peddler should be allowed to wreck his plans. Accordingly, Kit's men patrolled the trails — particularly to the south and west — and callously poured into the sand all the Spanish brandy they could find in the hands of any trader, white or Mexican. Bent kept his own kegs under lock and key and warned every man at the fort not to dare offer one drop to an Indian.

[1] Grinnell, *The Fighting Cheyennes*, chap. VI.

While his men were on this duty, Kit was much about the big camps along the river. On the north side were the Cheyennes and Arapahoes; on the south, just under the glaring white sandhills, stood the massed lodges of the Comanches, Apaches, Kiowas. It was a great time to talk over old scrapes, to find out what the Indians had to say about this fight and that, and how many were rubbed out the year it rained fire.

And it was in this big camp that Kit first saw Making-Out-the-Road, the Cheyenne girl. Her name evidently referred to some exploit of her ancestors, and meant making out or reading the sign of some trail or road made by the enemy. He first saw her riding about the camps on a spotted pony in the company of her chum, who sat astride behind her. The two girls were having a wonderful time among all those people. The chum was nothing much to look at, Kit decided. But Making-Out-the-Road had long, sleek blue-black hair, a fine oval face, small hands and feet, and that day wore a dress of flaming red strouding. She was vivacious, quick-tempered when her horse failed to obey her, and her temples were touched with vermillion. Kit decided it was about time he womaned.

In those days a white man was an important person in an Indian camp, and constantly under observation. Kit's attention to the Indian girl's beauty was immediately noticed. A young man, the cousin of the girl's chum, was one of her admirers. He immediately began to talk, and would announce in the girl's hearing that Kit was planning to 'play a trick' upon her. This amused the Indians, as Kit did not at that time understand their language, and was unaware of the joke.

Making-Out-the Road was embarrassed and became

exceedingly shy of Kit, while her father and the rest of the family became a little uneasy. Meanwhile the young cousin continued to warn the girl and amuse the camp. But no one can keep a secret in an Indian camp, and after a bit one of Kit's friends — a boy still in his teens — told him the state of the case.

At once Kit walked from the lodge where he was sitting to the tepee of Making-Out-the-Road. It was a warm day, and the lodge-skins had been rolled up on the shady side of the lodge to allow the breeze to blow through. Inside, in the cool shade, lay the father and brothers of the girl, and she herself was sitting there embroidering a moccasin. She looked very pretty, neat and clean, as she sat there — very properly — her small yellow moccasins together on the left. Before the tepee stood the chum's cousin, the young man who had caused all the trouble. He was talking to the girl, very earnestly, bragging about himself and what he was going to do the next time he went on the war-path against the Pawnees. She listened silently, while her father and brothers smoked.

Kit came up behind the young man, and, standing there, began to make sign talk to the girl and her family. Kit's gestures were a satirical comment on the young man's words and brags, and he kept it up a long five minutes. The people in the lodge entered into the joke with delight, watching Kit's hands, but never relaxing the gravity of their faces. They did not let on, and the young man remained there making a fool of himself while Kit called him everything ridiculous in the varied vocabulary of the sign talk. Finally the father smiled, and the girl laughed outright. The young man turned around, saw what was going on, and, muffling his head in his buffalo robe, marched off to his own tepee. Kit went into the lodge and talked turkey to the girl's father.

Making-Out-the-Road was a different proposition from the Arapaho girl. The Cheyennes were not the accommodating, gentle, artistic folk with whom Kit had been associated in the Arapaho camps. They were proud, touchy, undisciplined, quick to anger, and Making-Out-the-Road looked with disdain upon the Arapaho women. In those days the Cheyennes regarded the Arapahoes much as white men regard negroes, since the Arapahoes were few in number. And so Kit found that Waa-nibe's successor was something of a problem.

To begin with, she had the firm conviction that all white men were rich beyond the dreams of Indian avarice. There warn't enough vermillion, nor beads, nor scarlet cloth in Bent's big lodge for her. Traps wouldn't buy all the fofurraw she craved. And she had an insatiable appetite for the society of her friends and relatives — including the young cousin of her chum. Now Kit's business kept him moving and on the trail most of the year, and she had no mind for such traipsing up and down the land, worrying with pack-mules and travois. She much preferred putting on her beads and paint and red dress to sit by the fire with other young matrons, playing the hand game or tossing dice in a basket. She loved to dance. And though she could embroider prettily enough, she had the smallest kind of heart for the back-breaking labor of dressing buffalo robes. At least, that was Kit's side of the story. He had bargained for a helpmeet.

Nor did she take kindly to Adeline, especially after

she lost her own baby before it was a month old. She became hard to manage. And one day — when Kit told her it was time to move — she stubbornly refused to take down the tepee. By that time Kit had some command of the Cheyenne tongue, and attempted to argue the matter, making allowance for her quick temper. The Indians all maintain that Kit never beat his wife, never gave her a lodge-poling as did other trappers. But this time he was exasperated, and went into the lodge to let her understand who was boss. Not very long after he came out again, and after him came a cloud of saddlery, possible sacks, traps, lariats, blankets, clothing — everything he owned. The whole camp looked on. 'Thunder strike you!' she was screaming. 'I've thrown you away! You shall never sleep in my lodge again! *No-het'-to!* (That's that!)'

Well, after that Kit quit her.

Another trapper might have swapped her to some man for a saddle or a plug of tobacco. But Kit had been fond of her, and simply rode away.

Making-Out-the-Road did not wait to be informed of his intentions. As soon as he was gone, she went off with the young cousin of her chum. By all accounts, she made this young Indian an excellent wife. Several of her descendants by this second marriage are still living in Oklahoma.[1]

During Kit's temporary residence in the Cheyenne woman's lodge, the Carson Men gained an odd recruit in the person of a Cheyenne orphan boy — a distant connection of Making-Out-the-Road. He was marked by scrofula, and is said to have been very light in color

[1] *Vide* Grinnell, *Bent's Old Fort and Its Builders*, Kansas Historical Society Collections, vol. xv, p. 37.

for a Cheyenne. He took a violent fancy to Kit, and nothing could drive him away. At first the men tried to get rid of him and made game of him in various ways. He was fond of liquor and delighted them with his antics when drunk. Kit found him useful as a runner. But when sent on an errand he had a bad habit of borrowing somebody's horse without bothering to ask for it, and at last they gave him a pony of his own. At night this boy would turn up a kettle and beat on it with a stick and 'sing Injun,' making a most ungodly noise until some one shut him up. But the loveliest thing about this boy was his lovely name — Buffalo Chip — a name which gave rise to endless puns, jokes, and innuendoes among the mountaineers.

One day Buffalo Chip was sent to the fort for some powder and ball. He did not return. Afterward, it was discovered that he had there seen a girl he liked the looks of, had given the furs with which he had been intrusted to the girl's brother, and had married her the same day. The Carson Men probably considered that it was worth the furs to get rid of their little *Bois de Vache*.

In the spring of 1842, following his summary divorce of — or by — the Cheyenne girl, Kit decided to take Adeline to the settlements in Missouri. It was clear that a tepee was no place to bring up the child of a man of Kit Carson's standing. She was entitled to advantages, and he could well afford to educate her. Bent paid him a dollar a day, and he retained a tithe of the income of his band of trappers. He was successful now, and a vacation would do him good. He arranged to go east with Bent's annual caravan.

In the years following the peace treaty, Bent and St.

Vrain did a good business. The papers in Missouri constantly report the coming of their caravans 'with 15,000 buffalo robes ... and a considerable amount of furs' ... '36 days on the trip; brings a large lot of buffalo robes and furs.' ... 'Bent and Company will soon arriv: at St. Louis with upwards of 1100 packs of buffalo robes and 2–3000 lbs. of beaver.' ... 'Part of Bent's and St. Vrain's traders arrived with 283 packs of buffalo robes, 30 packs of beaver, 12 casks of tongues, 1 pack of deerskins. ...'

Sixteen years had passed since Kit left Franklin. He was now thirty-three years old, with the best — if not the happiest — years of his life ahead. Now he could ride his own horse with silver-mounted bridle. No more teaming, no more driving the cavvy. Even the end of the beaver trade had not downed him. He placed Adeline, now going on seven years, in Bent's Dearborn carriage, and, riding alongside, put out on the long trail back to the States.

The train set out in April, passing along the north bank of the sprawling Arkansas, and with it went a horde of Cheyennes — tepees, dogs, kettles, packmules, and all — heading for the buffalo range. Sol Silver waved his old wool hat from under the bastions of the big lodge, and prepared to lead the Carson Men away to run meat in Kit's absence.

The caravan passed down the stream, past the creek where Kit had fought the Crows ten years before, past the mouth of the River of Lost Souls, the Purgatoire, past Bent's Log Houses and through the Big Timbers, fording Sand Creek, crossing Big Salt Bottom. On they went, to Pretty Encampment, Chouteau's Island, Cimarron Crossing, the Caches, the treeless

Coon Creeks, Pawnee Fork, Ash Creek, Walnut Creek. There the Cheyennes left them, and the train pulled slowly on alone to Cow Creek, Little Arkansas, Council Grove, Big John Spring — Independence!

And Franklin — where was Franklin? A hundred and fifty miles down-river, they told him; a hundred and fifty miles inside the new frontier. Putting his daughter in the hands of relatives, he went back to have a look at Franklin and the principal streets eighty-two and a half feet wide. But that square containing two acres had been washed into the muddy Missouri; the 'agreeable and polished society,' the 'business and importance,' had vanished. Even the very soil was gone: only the graveyard on the hill remained.

A few days in St. Louis at the Rocky Mountain House; a few days greeting old friends and numerous relatives; then Kit turned his toes westward again, and stood at the rail of a steamer bound for the upper Missouri.

On board that steamer was Lieutenant John Charles Frémont, starting on his first expedition to the Rocky Mountains, and in need of a guide. Drips, whom he had hoped to employ, could not be found. Kit offered his services.

Frémont cautiously questioned the inoffensive little man. 'What experience have you had? Do you know the mountains?'

Kit spat into the boiling river water. 'I reckon so. A ten-prong buck warn't done suckin' when I last sot on a cheer!'

Frémont said he would make inquiries, did so, and promptly engaged Kit at a hundred dollars a month — just three times what Bent, the Yankee trader, was paying. Good-bye Bent and running meat. The Oregon Trail, South Pass, Fort Laramie, the Rockies — Kit knew them well.

Ten years in the mountains had made Kit brave and decisive. The four years on the plains had made him bolder and more humane. But now he was to come into close contact for the first time with men of wide culture and experience. He was to feel his lack of 'book l'arnin'.' He was putting out on a new trail leading to national and world-wide fame.

Kit used to say he owed more to Frémont than to any living man. Unquestionably, he did. Frémont was just the person to bring to light the excellence of the unassuming Kit, just the person to advertise him to the reading public and thereby serve his own and Senator Benton's aims. For — as with the Homeric epics, the Arthurian legends, the poems of Virgil — the romance of Kit Carson's adventures was partly moulded for a political purpose. Kit was to be the hero who personified American enterprise in the Far West — the banner which was to wave the pioneers forward into the Great American Desert.

Senator Benton and his party would never rest until the United States extended to the waters of the Pacific.

IV
PATHFINDER

CHAPTER XVI
JOSEFA

THREE wives within eight years simply looks like carelessness nowadays. But that was Kit Carson's record. Of course, under the rough conditions of pioneer life, women were soon knocked to pieces. Birth control was unheard of, and women did their full share of the back-breaking labor in camp and on the trail. They were helpmeets, sure enough. A man without a wife was a pitiable object in those days.

Then, too, it must not be forgotten that Kit belonged to that hardy race of mountaineers who came to America at a time when Henry VIII was fresh in the memories of men. That royal example must have counted for something with the pioneers. Some of the mountain men readily accepted the Indian custom of polygamy, though never on the magnificent scale of Brigham Young, the indomitable. Kit Carson, so far as is known, never had at one time more than one wife — certainly not more than one Cheyenne! But he was not a man who could long remain unwomaned.

There was a reason for his taking Adeline to Missouri, a reason for his jumping at Frémont's offer of a hundred dollars a month. This was the Señorita Josefa Jaramillo, daughter of Don Francisco Jaramillo and Maria Polonia Vigil, the belle of Taos, a girl just fifteen years old.

She was a beauty of that dark, precocious type found among the Southern peoples, a striking brunette, one of those rare women who have the gift to sparkle in repose. 'Her beauty was of the haughty, heart-breaking kind — such as would lead a man with the glance of the eye to risk his life for one smile,' says Garrard, who saw her four years later in the witness box at Taos, and comments upon the unusual refinement of her dress and manner. He adds, 'I could not but desire her acquaintance.'[1] Neither could Kit, though he was old enough to be her father.

She was vivid, alive in every fiber, and Kit first took note of her in a dress of that glorious yellow which appears to us in dreams as the color of pure happiness. There is no doubt that in her Kit found his match — a mate as passionate and quick and intelligent as himself. It was a love match — a rather rare thing among the Mexicans in those days, when parents consulted their own convenience in family alliances, and the price demanded by the *padres* for a marriage was extortionate.

Not that Kit did not do well in this marriage. The Jaramillo family was among the best in Taos. Charles Bent had married Josefa's elder sister, and the Vigils, the family of the girl's mother, had held high office in Santa Fe. Kit could look higher than an Indian woman now. Taos was his home. Probably it never even occurred to him that he might bring out a girl from Missouri. If he ever did 'have the idy,' it was too late when he started for Independence with the caravan. He had already fallen in love with Josefa.

A great deal has been made of Carson's services to

[1] Garrard, *Wah-To-Yah and the Taos Trail*, chap. xv.

the United States in piloting Frémont around the ranges, as though Kit could foresee the vast changes which would follow those explorations — so-called. But the fact is, Kit went with Frémont because he loved Josefa and wanted to better himself. Like most people who do things in the world of affairs, he was moved by no grand schemes or high-falutin' sense of service or honor, but simply set his heart on a woman and a little money. The honor — if it should come — would be all velvet.

Frémont's party, which consisted mostly of Frenchmen recruited in St. Louis, started from Chouteau's Landing near the mouth of the Kansas River. Lucien Maxwell, Kit's old friend and formerly a trader among the Arapahoes, was engaged as hunter. Preuss, a topographer, accompanied the party. Besides these, there were two boys: Henry Brant, nineteen, and Randolph Benton, son of the Senator, who was sent along, presumably, in order to convince people in the settlements that even a boy could traverse the prairies. Randolph was twelve years old. Kit was guide.

The party were all armed and mounted. In addition to the horsemen there were eight carts, one of which contained a fantastic contrivance which aroused inextinguishable laughter among the mountain men — an India-rubber boat! The darned thing came near killing Kit Carson. What an end for the Hero of the Plains — death dealt by an India-rubber boat! Such fofurraw disgusted Kit. Frémont had a lot to learn about the ways of the West.

But then, Frémont was a military man. And long after his day, military men pursued Plains Indians with howitzers and a pontoon train!

The trouble with the India-rubber boat began at once — at the ford of the Kansas River. There Basil Lajeunesse towed it over, swimming, with the end of a cord in his teeth. They got over most of the carts, but at last the load capsized. 'Carts, barrels, boxes, and bales, were in a moment floating down the current; but all the men who were on shore jumped into the water, without stopping to think if they could swim, and almost everything — even heavy articles, such as guns and lead — was recovered.'[1] Even the men unable to swim were saved.

They lost their sugar, their coffee — nearly all of it — and Kit and Maxwell, upon whom, of course, most of the labor fell, were compelled to lie in bed for a day after. To compensate for this unpleasantness, Frémont bought a milch cow and a bag of coffee from some Indians, and the party moved on. Frémont served breakfast before starting in the morning — a luxury to which the mountain men were quite unused. Barring the rubber boat, the expedition was rather a holiday to men like Kit and Maxwell.

Frémont, no doubt consulting the wishes of young Benton, put the two boys on the roster of the guard on the same tour of duty with Kit, and Kit became the Nestor of the Prairies to those eager youngsters. He taught them how to make their beds, how to build a fire, how to dig a *cache*, how to read sign, and told them stories of Injun scrapes: about the man who, scalped alive, went on fighting with his pistol in one hand and the other hand holding up the sagging skin of his forehead, so that he could see to shoot . . . about the Sioux scouts who disguised themselves as white wolves on

[1] Frémont, *Report*, p. 174.

moonlight nights and scampered about his camp, pretending to snap at his dogs, and imitating the click of the wolf-teeth with a couple of buffalo bones held in the hand. He told them how the Spaniards dried meat in long strips and sold it by the yard, and why the streams and islands they passed bore the names of trappers rubbed out there by the Injuns.

He told them how Brady's Island was named for a man who had had a quarrel with a companion, and was left with this man to guard a boat. On the return of the party, Brady was found dead, shot by his own gun — accidentally, his companion said.

He told them how Gonneville, a trapper, had been killed by the random shot of a Yankton Sioux after a brave fight, and how he had left his name to Gonneville Creek. He pointed out Court House Rock, Chimney Rock, and Scott's Bluffs, and told the melancholy story connected with them. How Scott, too ill to move, had been deserted by his starving comrades; how he had waited in vain for their return, and had then set out to *crawl* to a camp of white men believed to be down-river. How, the next spring, Scott's skeleton had been found *forty miles* from the spot where he was left to die!

Young Benton learned to use his saddle for a pillow and a barricade for his head at night, with his pistols laid above it. He was taught to spread his buffalo robe on high ground and with a skinning-knife to ditch his bed all round against the rains. He learned how to *cache* his trusty rifle, loaded and ready, under the waterproof Navajo blanket which covered him.

Kit taught him the folly of sitting in the firelight at night, when some young Indian, eager to strike a *coup* and careless who died for it, would be likely to send an arrow into him from the darkness. Randolph heard that the best way to extract an arrow was to push it on through and then, having cut off the iron-barbed head, to withdraw the naked shaft; or — if pushing the arrow through seemed dangerous to life — to wait until the blood had softened the sinews fastening the head to the shaft, when the shaft might easily be drawn away, leaving the head in the wound. Then the head might, perhaps, be butchered out with a skinning-knife.

Kit taught him how to strike fire into a bit of dry punk with flint and steel, how to nest the glowing tinder in a handful of dry grass to be waved in the air until the flame came, when it would kindle the dry twigs forming the foundation of the fire.

And by example and precept the boy learned from Kit that skill and caution and courage would conquer most difficulties, and that — if they did not — it was no use to complain.

At the end of the fortnight they had the first Indian scare, and Kit, 'springing upon one of the hunting horses, crossed the river, and galloped off into the opposite prairies, to obtain some certain intelligence of their movements.

'Mounted on a fine horse, without a saddle, and scouring bareheaded over the prairies, Kit was one of the finest pictures of a horseman I have ever seen. A short time enabled him to discover that the Indian war-party of twenty-seven consisted of six elk, who had been gazing curiously at our caravan as it passed by, and were now scampering off at full speed,' says Frémont.

On the way up the Platte they encountered three Cheyennes — two men and a boy of thirteen — just returning from a fruitless raid upon the Pawnee horse-herds. They reported that the Pawnees were cowards, who shut up their horses in their lodges at night. They had only lances, and rode sorry wild horses taken on the Arkansas River. Soon after, the party saw their first buffalo. 'It was the early part of the day, when the herds were feeding; and everywhere they were in motion. Here and there a huge old bull was rolling in the grass, and clouds of dust rose from various parts of the bands, each the scene of some obstinate fight.' It was the rutting season. 'Indians and buffalo make the poetry and life of the prairie,' adds Frémont. 'Three cows were killed to-day. Kit Carson had shot one, and was continuing the chase in the midst of another herd, when his horse fell headlong, but sprang up and joined the flying herd. Though considerably hurt, he had the good fortune to break no bones; and Maxwell, who was mounted on a fleet hunter, captured the runaway after a hard chase. He was on the point of shooting him, to avoid the loss of his bridle (a handsomely mounted Spanish one) when he found that his horse was able to come up with him.'

Next day they had another chase. 'The wind was favorable; the coolness of the morning invited to exercise; the ground was apparently good, and the distance across the prairie (two or three miles) gave us a fine opportunity to charge them before they could get among the river hills. . . . They were now somewhat less than half a mile distant, and we rode easily along until within about three hundred yards, when a sudden agitation, a wavering in the band, and a galloping to and fro of some which were scattered along the skirts, gave us the intimation that we were discovered.... The front of the mass was already in rapid motion to the hills, and in a few seconds the movement had communicated itself to the whole herd.

'A crowd of bulls, as usual, brought up the rear, and every now and then some of them faced about, and then dashed on after the band a short distance, and turned and looked again, as if more than half inclined to stand and fight ... the rout was universal ... we gave the usual shout, and broke into the herd. We entered on the side, the mass giving way in every direction in their heedless course. Many of the bulls, less active and less fleet than the cows, paying no attention to the ground, and occupied solely with the hunter, were precipitated to the earth with great force, rolling over and over with the violence of the shock, and hardly distinguishable in the dust. We separated on entering, each singling out his game.

'My horse was a trained hunter, famous in the West under the name of Proveau, and, with his eyes flashing, and the foam flying from his mouth, sprang on after the cow like a tiger. In a few moments he brought me alongside of her, and, rising in the stirrups, I fired at the distance of a yard, the ball entering at the termination of the long hair, and passing near the heart. She fell headlong at the report of the gun, and, checking my horse, I looked around for my companions. At a little distance, Kit was on the ground, engaged in tying his horse to the horns of a cow which he was preparing to cut up. Among the scattered bands, at some distance below, I caught a glimpse of Maxwell; and while I was looking, a light wreath of white smoke curled away from his gun, from which I was too far to hear the report. Nearer, and between me and the hills, towards which they were directing their course, was the body of the herd, and, giving my horse the rein, we dashed after them. . . .'

Well, with a press-agent like that, is it any wonder that the trail to Oregon was crowded within a few years? Kit Carson no longer need concern himself with fame: Frémont will take care of that for him. And indeed, Frémont took such excellent care of it that Josefa and the dollars — the objects of Kit's solicitude — came near being crowded out of his later life. Frémont and those who followed him kept Kit so busy that he rarely found time to spend with his handsome wife in Taos.

Running meat was of no interest to Kit Carson. The thing which he remembered about this first expedition with Frémont was that he came near losing his hair under the orders of that impetuous leader.

When they got to Fort Laramie, every one agreed that to go farther was to court almost certain destruction. The Snakes and trappers some time before had killed some Sioux, and now the Sioux — more than a thousand lodges, it was said — were on the war-path, determined to have revenge.

Bordeaux, the trader at the fort, and his two clerks, Galpin and Kellogg, sat among their bales of buffalo robes, calicoes, blankets, lead, powder, guns, glass beads, looking-glasses, rings, vermilion, tobacco, and kegs of fiery liquor, and dilated upon the danger of proceeding. The gateway of the fort was the only cool place in the establishment, and there the traders remained.

Old Frapp, they said, had been wiped out, and with him four of his men on Snake River. The Indians had lost ten men, and were out for hair. Since then two other small parties had been murdered not far away. The emigrants who had preceded Frémont had become so discouraged that they sold their oxen and scurried over to Fort Hall under White Head's guidance. The Sioux, Cheyenne, and Gros Ventres were united to make an attack upon a band of trappers in the Crow country, the whole region was swarming with roving war-parties, and Frémont was not surprised that his men became alarmed. 'Carson, one of the best and most experienced mountaineers, fully supported the opinion given by Bridger of the dangerous state of the country, and openly expressed his conviction that we could not escape without some sharp encounters with the Indians. In addition to this, he made his will; and among the circumstances which were constantly occurring to increase their alarm, this was the most unfortunate; and I found that a number of my party had become so much intimidated, that they had requested to be discharged at this place,' says Frémont. When Kit Carson made his will, small wonder that the Frenchmen got scared.

But Frémont did not waver. If his rhetoric was gorgeously Latin, so were his dash and courage. Kit found that a contract with Frémont meant more than an easy hundred dollars a month.

Dramatic as usual, Frémont lined up his men, warned them of the dangers, declared that 'these were the dangers of everyday occurrence, and to be expected in the ordinary course of their service. They had heard of the unsettled condition of the country before leaving ver and Canty. Silver told my father that when Kit got back to Taos there was a great celebration, and that he (Sol) got drunk and fell off his horse, and was laid up for three days as a result of this spree. This was in January, 1843.

On February 6th, 'Cristover Carson & Maria Josefa Jaramillo' were united by the parish priest, Antonio José Martínez. The witnesses were George Bent, younger brother of William and Charles, and three Mexicans. The crumbling towers of the old church echoed to the clank of the flat-toned bells, the trappers whooped and drank at the fandango which followed, and Kit Carson, newly converted to the Roman Catholic Faith, led his handsome bride back to the house which still stands near the Taos plaza.

That house has become a place of pilgrimage. The Masons — for Kit was a Mason — have bought it and restored the building. Tourists throng its long *portal* every summer. The lady who lives there answers your knock with a certain justifiable hauteur; for the tourists, she says, invariably ask her, 'Are you Mrs. Kit Carson?'

Josefa was born in 1828!

work, and, although all three of them had lived in Indian lodges for years, not one could recall the order or manner of placing the poles so that the cover would fit. They argued and sweat over it most of the afternoon, and the result of their labors looked like a wrecked umbrella.

Then Bisonette and his Indian wife turned up. She laughed at the crazy lodge. Then, with an air of great superiority to mere men, she went about her work, and in fifteen minutes had the big lodge trim and taut and a fire under the kettle inside. For some days she pitched and struck the tent, until Frémont's Frenchmen had mastered the simple trick.[1] After that, the party slept in comfort.

The Sioux war-parties failed to oppose the march, and the trip was uneventful except for the wrecking of the rubber boat, which almost drowned all the men aboard in the Red Narrows of the Platte. Frémont completed his survey of the South Pass and the Sweetwater Trail, and, having got back to Fort Laramie, proceeded to go home and get his Report printed. Senator Benton made a great to-do over the expedition, and the first blow was struck at the old myth and bugaboo — the Great American Desert.

Carson took his pay and led his Carson Men back home to Bent's Fort and Taos — for of course some of his faithful band accompanied him. Sabin[2] quotes Oliver Wiggins as saying that Kit put no faith in Frémont's company, and with his usual foresight 'preferred to have his men within call in case of trouble.' Among those who responded to his call were Sol Sil-

---

[1] *Vide American Anthropologist*, January–March, 1927, pp. 87 ff.
[2] Sabin, *Kit Carson Days*, p. 216.

ing things for their effect as propaganda. While Frémont was stunning the Sioux chiefs with his oratory, Kit was waiting to see his commander about a matter he considered of more vital importance.

Like any thorough outdoor man, Kit was keenly interested in his own comfort. Danger might — or might not — come, but daily comfort was essential if life in the wilds was to be worth living. When the council was ended, Frémont found Kit waiting before his tent — a matter of importance, he said. Frémont was ready to hear him, probably expecting some deep-laid plot of strategy against the foe. But Kit merely requested that Frémont buy an Indian tepee. Kit and Maxwell were sick and tired of the ridiculous wall tents supplied by the War Department. Kit pointed out that the tepee was staunch in the heaviest winds, that it shed water well, that it could be heated by a fire inside, that the sides could be raised for coolness in hot weather, and that you never found mosquitoes in a tepee. The wall tents were hard to pitch, cold in winter, hot in summer, unventilated, and would not stand up against the western winds. In short, the wall tent was made to sell and the tepee to live in.

Frémont at once saw the justice of these opinions. He discarded his clumsy imitations of a house without windows, and bought 'a tolerably large lodge, about eighteen feet in diameter and twenty feet in height.' Then he packed it over to his camp and ordered Kit to pitch it.

That stumped Kit and Maxwell, and all the Carson Men who were along. Sol Silver remembered that Kit and he and Maxwell wrestled with the poles and covering for nearly an hour. Pitching the tepee was woman's

St. Louis, and therefore could not make it a reason for breaking their engagements. Still, I was unwilling to take with me, on a service of some certain danger, men on whom I could not rely; and as I had understood that there were among them some who were disposed to cowardice, and anxious to return, they had but to come forward at once, and state their desire, and they would be discharged with the amount due them.' One man came forward. Frémont exposed him to the ridicule of the rest, and let him go.

Having gained his point, Frémont very sensibly left young Benton and Brant at the fort. The men missed Randolph, their *petit garçon*, who had been the life of the party.

At once Bisonette and the Sioux chiefs began to dissuade the party from going on. Otter Hat, Arrow Breaker, Black Night, and Bull Tail insisted that their young men could not be controlled. Frémont demanded that one of them go with him to act as peacemaker. They declined. So Frémont told them, 'We have thrown away our bodies and will not turn back . . . you did not see the rifles that my young men carry . . . you are many and may kill us all; but there will be much crying in your villages. . . . Do you think that our great chief will let his soldiers die, and forget to cover their graves?' The Indians said they would send a man along.

All these heroics hardly interested Kit Carson. Life was very precious to Kit just then because of Josefa, and very naturally he made his will, as soldiers and trappers have always been accustomed to do before a battle. But Frémont's excitement over the matter must have puzzled Kit, who had never been used to do-

CHAPTER XVII
THE SPANISH TRAIL

No sooner had Lieutenant Frémont read proof on his first Report than Senator Benton engineered a second expedition. This time the purpose was ostensibly to 'connect the reconnaissance of 1842 with the surveys of Commander Wilkes on the coast of the Pacific Ocean.' The Senator, having now demonstrated that the trail to Oregon was easy and safe, followed up by arousing the land-greed of the American public. Frémont was to go on to Oregon and North California, and to return with a glowing report of the country beyond the Sierras. He set out late in May, 1843, from Westport Landing, and was gone more than a year.

Frémont hastened his departure, as he had private information — from his wife (daughter of Senator Benton) — that the War Department had sent orders for his return, demanding an explanation in writing of his conduct in making a howitzer part of his equipment on a peaceful reconnaissance. Frémont and Benton cared little whether their joint expedition stirred up war with Mexico or not. But the War Department feared trouble with England. And so Frémont made haste to get out of range of the big guns of the War Department.

This time he took along a more ambitious staff. White Head Fitzpatrick was guide. Gilpin, a newspaper man of Benton's party, went along. A negro freedman, a Prussian gunner — for the howitzer — and several other Americans, besides an enlarged roster of Frenchmen from St. Louis. Maxwell joined the party on the way out. The route was through Kansas to the South Platte in Colorado and on to Fort St. Vrain, one of Bent's posts. From that post Frémont sent to Taos for mules. At Pueblo, Colorado, they found Kit Carson, who was persuaded to join the expedition.

While Frémont was advancing through Kansas to the Rockies, Kit had been riding in the other direction as hunter for Bent's caravan. His honeymoon had lasted but a little more than two months. At Walnut Creek Kit met the dragoons under Captain Philip St. George Cooke. A rich Mexican caravan belonging to Armijo, then Governor of New Mexico, was rolling along behind, and the dragoons had been sent as an escort to protect the caravan from the Texans, who were waiting on the border to intercept it. For the Texans had good reason to hate Armijo, who had treated their prisoners (men of the Texan Santa Fe Expedition) so brutally. Beyond the ford of the Arkansas River (the international boundary) Cooke could not go, and the Mexicans were very much afraid. They offered Kit three hundred dollars if he would carry a letter to Armijo in Santa Fe, asking for a guard through Mexican territory. The offer was promptly accepted, and Kit, taking Dick Owens along, galloped back to Bent's Old Fort. There he heard that the Utes were out on the trail to Taos. But Kit wanted that three hundred, wanted to see his wife, and Bent let him have the fastest horse in the corral. Leading this animal and riding another, Kit found the Ute village before the Indians saw him. He hid until dark. Then rode on and reached Taos in safety. A man with a led horse can, of course, choose his trail and avoid making much sign. The letter was sent on to Santa Fe, and four days later Kit got the answer to carry back.

Sol Silver rode with Kit. On the way they encountered a band of Ute warriors, who at once rode up, evidently with hostile intentions. Sol urged Kit to ride on with the letter, for it was evident that none of the Indians had a horse that could overtake Bent's fine animal. Sol told Kit that he could manage the Indians, and that they would not be likely to hurt him in any case. Kit's errand was a responsible one, and, if the letter miscarried, might mean the death of many men. At first thought, he agreed with Sol, mounted the fast horse, and prepared to make a dash to safety.

But as Kit saw the war party coming, he changed his mind. It would be cowardly, he saw, to risk his companion's life, leaving him to the uncertain mercies of the savages. He got off his horse and took his rifle from its case. 'I reckon I'll stay here,' said Kit. 'If they rub us out, we'll go under together.'

The chief rode up to Kit, offered his hand. But when Kit reached out to take it, the Indian seized Kit's rifle and tried to wrench it away from him. This was a favorite trick of Indians on the plains. If the white man yielded his gun, he was taken without risk. If he resisted, it gave an excuse for killing him. But Kit was not the greenhorn the chief took him for. They wrestled for the gun, and Kit made the man let go. By that time the rest of the war-party had ridden up. They rode around Kit and Sol, who stood back to back. The Indians talked in loud, threatening tones, and brandished their guns, shaking the priming into the pans, hoping that the two men would be frightened or change their positions so as to give them an opening. But Kit and Sol kept their rifles ready, telling the Utes to take care and keep off, if they did not wish to lose two warriors. After a while the Utes decided that nothing but hard knocks could be got from two such men, and rode away. The rest of the trip passed without any trouble.

At Bent's Fort Kit learned that the Texans had been caught and disarmed by Captain Cooke's dragoons. Hearing that Frémont was not far off, Kit rode over to see him. He had nothing to do now, for Bent's caravan was far out of reach. Moreover, he had seen his wife and had made a handsome sum within a few days. Frémont, realizing the value of Kit's services, at once engaged him.

Immediately Frémont sent White Head along with the baggage train, and set out with Kit and fifteen men for Thompson's Fork, on by the Cache la Poudre, Laramie Plains, and North Platte to Sweetwater and Devil's Gate. From there Frémont turned aside to visit the Great Salt Lake, taking his rubber boat with him. On September 6th, he saw the lake. Three days later he and Kit and three others got into the boat and made for an island; it was Kit's first attempt at navigation. He was excited, his head full of the yarns about giants and monsters, and, when white-caps appeared at a distance, earnestly urged Frémont to take a look at them through his glass.

Leaving Salt Lake, the party found White Head at Fort Hall, and all together set out to the mouth of the Columbia. Here Kit's men turned back, and Frémont, taking Kit and his own men, pushed on to the coast. This was the limit of Frémont's expedition as authorized by the War Department, but it would be hard to go back in winter, and anyway he wanted to go south. He went south. The Great Basin nearly finished the command, the snow on the Sierras accounted for half the horses and the howitzer, but at last they arrived at Sutter's Fort in the valley of the Sacramento in California. They went on up the San Joaquin, and Kit felt more at home. For he had been in this region with Ewing Young. Striking south and east, the party crossed a pass into the Mohave Desert, and on to the Spanish Trail.

Here occurred the incident which first made a national hero of Kit Carson. Two Mexicans appeared suddenly in camp one afternoon — a man and a boy. 'The name of the man was Andreas Fuentes; and that of the boy (a handsome lad, eleven years old) Pablo Hernandez. They belonged to a party consisting of six persons, the remaining four being the wife of Fuentes, the father and mother of Pablo, and Santiago Giacome, a resident of New Mexico. With a cavalcade of about thirty horses, they had come out from Pueblo de los Angeles . . . in advance of the great caravan, in order to travel more at leisure, and obtain better grass. They halted. . . . Several Indians were soon discovered lurking about the camp, who, in a day or two after, came in, and, after behaving in a very friendly manner, took their leave, without awakening any suspicions. . . . In a few days afterwards, suddenly a party of about one hundred Indians appeared in sight, advancing towards the camp. It was too late . . . the Indians charged down into their camp, shouting as they advanced, and discharging flights of arrows. Pablo and Fuentes were on horseguard at the time, and mounted, according to the custom of the country. One of the principal objects of the Indians was to get possession of the horses, and part of them immediately surrounded the band; but in obedience to the shouts of Giacome, Fuentes drove the animals over and through the assailants, in spite of their arrows; and, abandoning the rest to their fate, carried them off at speed across the plain. Knowing that they would be pursued by the Indians, without making any halt except to shift their saddles to other horses, they drove them for about sixty miles, and this morning left them at a watering place on the trail ... they hurried on, hoping to meet the Spanish caravan, when they discovered my camp ... I ... promised them such aid as circumstances might put it in my power to give.' So Frémont.

In spite of the hard work of the two Mexicans, when Frémont's men reached the spring where the horses had been left, they found that the Indians had run them off. Carson and Godey volunteered to go with the Mexican, and the three set out. ... In the evening Fuentes returned, his horse having given out. Kit and Godey went on alone, following the trail of the Indians. The sign was plain all the way, and led them into the mountains.

The reading of sign was a matter of trained observation and common sense. All mountain men were more or less expert at it. Of course, constant practice was necessary. The Australian blacks, who live upon mice and such small deer, are famous for their skill in tracking. No wonder: the track of a mouse is very fine script indeed. In comparison, a man's trail is written in letters a foot high.

Nowadays it is the fashion to laugh at Sherlock

Holmes and his magnifying glass. But for all that, it is certain that nobody can put his foot down without some disturbance of the surface where he sets it, and those traces can be seen and interpreted by one who knows what to look for and what it means. Let near-sighted, unobservant people laugh. Mountain men made a living following just such clues.

Rain, wind, snow, the tracks of passing buffalo — all these might obliterate or confuse the trail, making it hard to pick up and follow. But even under the most unfavorable conditions, some traces would remain. Kit knew that the stolen horses had passed only a few hours before, and since they were herded at a run, they had loped along together, making a plain track. In order to travel rapidly, Kit kept his eyes well ahead, since in that way he could see a series of tracks, whereas, if he had looked directly down, he would have seen — or missed — the single track at his feet. When he lost the trail, he had only to circle the last visible track until he found it again. Failing that, his common sense would tell him what the Indians were likely to do at such a place. He and Godey rode hard, not merely because night was coming on and they wished to overtake the Indians, but because the faster they traveled, the shorter the trail. Also, a fresh trail was always easier to follow than a stale one.

Kit was among the best trailers on the plains, though perhaps not the equal of the best Indian specialists in that field. He could determine at a glance the species, age, sex, gait, and sometimes the state of mind of whatever animal had made the tracks he found. He could recognize the hoofprints of his own horses and those of his constant companions. He could tell the tribe of the

Indian whose moccasin track was under his observation. He knew whether a trail had been made before or after dew-fall, and could tell from the warmth and consistency of the horse-droppings how long before the stolen herd had passed. The game was not disturbed. The birds gave no warning cries. Kit knew that his enemies were not stirring in his immediate neighborhood.

The trail through the mountain defiles was easy to follow, as there were few places where the Indians could have turned off. Even darkness did not stop Kit. He rode on, occasionally stopping to dismount and feel the ground with his hand, thus making sure that the inequalities were caused by the hooves of the stolen stock.

Finally, the men decided to snatch a little rest. They tied up their mounts and slept until nearly morning. Then, riding on, they soon discovered the missing herd, and, beyond the ridge, four lodges. The sun was rising, and the Indians, gathered about their fires, were feasting upon the flesh of one of the captured horses. Four others, partly cut up, lay near.

Kit and Godey left their horses, which were so used up that they could go no farther. Hiding them behind the rocks, they crawled forward, hoping to get the herd away before the alarm could be given. Of course, one of the animals snorted at the creeping figures, and the herd began to move off, alarmed by the snorting. The Indians at once jumped up, grabbed their arms, and Kit and Godey saw that they were discovered.

Said Kit, 'I reckon it's about time to charge, Godey.'

'You bet,' said Godey, and the two ran forward.

Kit fired, and one Indian dropped. Godey fired, but missed. Swearing at his failure, Godey stopped, jerked the plug from his powder-horn with his teeth, poured some of the powder into the muzzle of his gun, dropped in a ball, and without pausing to ram it home, fired again. The arrows flew like rain about him. But he dropped his target.

The Indians, alarmed by the dauntless attack, and probably supposing the two white men were the advance guard of a much larger party, promptly scattered, and began to run for the near-by hills. Kit reloaded, and, taking his stand above the camp on a little hill near by, kept guard while Godey got out his knife to scalp the two dead warriors. This arrangement was probably not due to any fastidiousness on Kit's part (though it was his boast that he never had knowingly scalped a man alive), but may have been prompted by the fact that Kit was a better shot than Godey, as just shown in the skirmish. A great deal has been said of the frontiersmen's unerring aim, and undoubtedly they were excellent marksmen, considering the kind of weapons they used. But Kit in his memoirs remarks upon the fact that, with three shots, he and Godey accounted for two savages: evidently he thought that mighty good shooting. It was. To-day, with our vastly improved arms, a soldier could not begin to carry the amount of lead that is fired in order to kill him.

Godey scalped his own man, but when he went to work on Kit's, the Indian came alive, letting out a hideous howl, while the blood streamed down from his skinned head. At the same time he loosed an arrow at his scalper, and Godey got it through his shirt collar. An old squaw, perhaps the Indian's mother, stopped and looked back from the side of the mountain she was climbing, shaking her fists, threatening and lamenting.

Godey was surprised, but the Indian fell back behind his cover of rocks, and Godey dashed in after him, and finished him off with his knife.

They found a boy, left by the fire, and tied him up. The prisoner, undismayed by the fate of his two companions, philosophically resumed his interrupted breakfast of boiled horse-head. Exploring the lodges, they found baskets containing fifty pairs of moccasins — a find which looked like pirate gold to the womanless trappers, who were always half barefoot. Rounding up the horses, they drove them back, riding new mounts in order to spare their own worn-out beasts.

On the way back they found the bodies of the two Mexican horseherders, both horribly mutilated. Afterward, the women were found, their bodies staked to the ground and similarly mangled.

Let Frémont tell it: 'In the afternoon ... a warwhoop was heard, such as Indians make when returning from a victorious enterprise; and soon Carson and Godey appeared, driving before them a band of horses, recognized by Fuentes to be part of those they had lost. Two bloody scalps, dangling from the end of Godey's gun, announced that they had overtaken the Indians as well as the horses. . . . The time, place, object, and numbers, considered, this expedition of Carson and Godey may be considered among the boldest and most disinterested which the annals of western adventure, so full of daring deeds, can present. Two men, in a savage desert, pursue day and night an unknown body of Indians into the defiles of an unknown mountain — attack them on sight, without counting numbers — and defeat them in an instant — and for what? To punish the robbers of the desert, and to avenge the wrongs of

Mexicans whom they did not know. I repeat: it was Carson and Godey who did this — the former an American, born in the Boonslick County of Missouri; the latter a Frenchman, born in St. Louis — and both trained in western enterprise from early life.'[1]

By Robidoux's fort on the Uintah, up the Yampah, and south through the Parks of Colorado, to Bent's Fort they came at last.

'As we emerged into view from the groves on the river, we were saluted with a display of the national flag and repeated discharges from the guns of the fort, where we were received by Mr. George Bent with a cordial welcome and a friendly hospitality.' That Fourth of July Bent's Old Fort kept open house to Frémont and his men. Charlotte did her best, and the negro servants were kept busy bringing ice for the juleps considered necessary to celebrate the occasion.

This was Kit's longest expedition, and he gladly left Frémont at Fort Bent and rode off home to Taos and Josefa.

Frémont went to St. Louis and home to get his Report into print. That Report determined the Mormons on their westward march, and many another American, young and old, began to long for the freedom of the plains and mountains, and to hope and plan to pull up stakes and put out on the trail to California.

Frémont's story of Kit and Godey got into the Eastern papers, and from that day Kit Carson's fame was genuine and growing wherever English print was read. To Carson, the incident was creditable, but not especially remarkable. He had been bred to just such adventures. But blazoned in Frémont's glowing style,

[1] Frémont, *Report*, p. 174.

it caught the fancy of the public here and abroad. For Frémont's Report had all the virtues of good advertising: it was arresting and interesting, and at the same time verified by its official publication and the seal of the United States upon its truth.

What had been vague and hearsay became authentic. The reading public became interested in the Far West. Immediately after the appearance of the Report, writers swarmed upon the plains, gleaning in the tracks of the Pathfinder: Parkman, Ruxton, Garrard, and all the rest. And all of them paid tribute to the mountain man Kit Carson, now suddenly sprung into popular favor as a romantic hero.

Now and again a voice was heard crying in the wilderness that Old Gabe or some other was as able and brave as Kit. But the public did not heed, and, as opportunity followed his rise into fame, Kit outdistanced all competitors for his place as the Hero of the Prairies.

He grew up with the country, and left them all behind.

CHAPTER XVIII
KLAMATH LAKE

THAT winter in Taos, Kit and Dick Owens decided that they would settle down and farm. Kit was fond of his wife. He had a family. She wanted him to stay near home. Kit's eyes had been opened to the changes rapidly taking place in the West. The trails to California and Oregon were crowded. Texas was soon to come into the Union. The Mormons were ready to take the trail for Utah, and the boundary between Oregon and Canada was about to be defined. Furs had failed. Indians were becoming so restless that trade with them was hazardous. Scouting with the Government expeditions was a business full of hardship and risk and long absence from home. Kit and Dick chose a place on the Cimarron River some fifty miles from Taos. There they broke the sod, put in their crops, and made some improvements.

Kit felt that he had rambled far enough. He knew all the country west of the Missouri from Chihuahua to Belly River, and for all his wanderings he had little to show except his well-deserved reputation as a first-class mountain man. His fame in the greater world of the East was unknown to him, and if he had been aware of it, Kit was the last man in the world to clown for what it would bring. He never tried to capitalize his reputation. He found the business of leading an honest, courageous, and successful life enough for him: fame and honor were by-products.

This strong, single-minded character was the greatest

factor in his success. Kit was not brilliant, not impressive, not a boaster. He first made sure he was right, and then went ahead without a scruple. Such a life was natural to him.

If he had been otherwise, he might not have lived to become famous. A fighter — particularly in a land where every stranger was a potential enemy — would not live long if he were beset by doubts, by scruples, by conflicting theories of conduct. While he was debating, the other fellow would take his hair. Nobody so much as got his fingers on Kit's scalplock.

Integrity meant courage and instant decision, and those meant success. For a man on the frontier honesty was not merely the best policy; it was the only policy, for it left the mind in a state of perfect freedom to act swiftly and effectively. Kit Carson always acted so.

We know that at thirty-two Kit had killed nineteen men.[1] His rifle of that date exists, with the brass tacks in the stock. During the years with Frémont (1842–47) he must have accounted for as many more. By his own account he engaged in thirteen fights during that period, some of them pitched battles with large bodies of Indians and Spaniards. And from other sources we know that he took part in minor scrapes — little skirmishes which he did not think it worth his while to mention.

Now, although Kit sometimes missed his target — as we know from two or three recorded instances — it isn't likely that he failed to 'bring' some enemy in each of those engagements, allowing for the fact that one of them has been described as a slaughter, 'a perfect butchery' of a camp of one thousand hostiles. In the

[1] Sabin, *Kit Carson Days*, note opposite p. 332.

fights where he opposed but one enemy, he was uniformly successful. It isn't likely that he failed to kill some one when targets were more numerous.

But these killings are not to be compared with those of the bad men of a later period on the Plains. Then, when danger from Indians had abated and the frontier was thronged by fugitives from justice, many such professional bravos appeared. Men like Billy the Kid, who lived to kill. Men like some of the marshals, who had the same spirit as Billy, but preferred to do their killing legally. . . . Kit Carson fought one duel, in an age when dueling was still a fashionable custom in high circles. All his other *coups* were counted in open warfare upon Indians or bandits or the soldiers of forces opposing the United States. He never picked a fight.

Kit entertained no silly twaddle regarding humanity or the rights of his professed enemies. He had been brought up to believe that every man should look out for himself. He was sometimes sorry that he had killed an enemy, but he was never in any doubt about the rightness of the action. As with buffalo, he killed no more than was necessary: all who knew him are agreed upon his mildness, gentleness, caution, and generally inoffensive manner. He did not undertake enterprises which he could not approve. Once having approved them, all who opposed him deserved what they got. And it was delivered immediately.

But now, on his farm, Kit is settled down — for four months. Then he got a message from Frémont reminding him of his promise to serve again if Frémont ever desired it. At that, Kit was plumb disgusted. But he never even thought of going back on his word. So he and Dick sold their farm for about half what it was worth, went to Bent's Fort, and joined the third and most dangerous of Frémont's expeditions.

The party left Bent's Fort in August, and 'took up' the Arkansas River, passing through the Bayou Salade to avoid the canyon, and striking the Arkansas again above it. On they went over the old trails to Grand River, White River, Green River, the Uintah, Provost Fork, Little Utah Lake, Great Salt Lake, on across the desert.

Kit and the other mountain men had never ventured into the desert at that point. And the terrified Indian guides insisted that there was no grass, no water — nothing. But Frémont was bound to cross just there, and nothing short of impossibility could stop him. He hated following old trails. He ordered Kit, Maxwell, Lajeunesse, and Archambeau (his best men) to go ahead and signal back when they reached water.

They set out over level barren country and went sixty miles to the mountains before they found water. They made a smoke to signal Frémont as agreed, and Archambeau set out to meet their commander. Frémont, having been on the lookout for the smoke signal with his glasses, had seen it and set out. He and Archambeau met in the midst of the desert, and a day later they reached Kit's camp, having lost only a few pack-mules.

Pushing on through the Basin in two parties, they crossed the Sierras into California and arrived at Sutter's Fort. Returning through the mountains, they failed to meet the second party, but had a skirmish with Indians, in which they killed five. Again they crossed the mountains, found their party, and set out for Monterey for supplies. But the Spanish General Castro ordered Frémont to leave the country at once.

The Spaniards had given no permission for this armed expedition into their territories. Castro threatened to drive them out.

Frémont made his camp on a hill, threw up fortifications, raised the American flag, and defied Castro to do his worst. The Spanish kept firing to scare Frémont, but their three hundred soldiers had no mind to attack forty mountain men, and at last the Americans got tired of waiting for the Spaniards to attack, and marched away. They stopped for a time at Peter Lawson's on the Sacramento River.

Already the war with Mexico was on, though Frémont and his men did not know it. But Senator Benton and his party had done what they could to make the war inevitable, and Frémont shared their hope and desire to annex the Mexican territories. His defiance of a perfectly proper order from the constituted authorities showed his purpose, and his raising the flag showed his hope. Probably the annexation was inevitable anyhow. But Frémont deserves credit of the expansionists for his courageous stand.

At Lawson's Kit became involved in a battle with one thousand hostile Indians, banded together to attack the whites. The campaign was a great success, ending in the destruction of their camp and a terrible butchery of the Indians. Afterward, Frémont's men set out up the Sacramento and, passing the snow-capped Shasta Butte, arrived on Klamath Lake. Their goal was the Columbia.

There two men overtook them, saying that Lieutenant Gillespie was not far behind with dispatches for Frémont. Gillespie's horses were used up, and he could not follow. As the Indians were all about and anything but friendly, Frémont decided to take ten picked men and travel the sixty miles at dawn in order to save Gillespie and his three companions. Frémont was also very impatient to read the dispatches.

Next day he made the journey, and reached Gillespie's camp before dark. Kit Carson, Dick Owens, Stepp, Godey, Basil Lajeunesse, four Delaware Indians and Denny, an Iowa half-breed, followed Frémont. They found Lieutenant Gillespie of the Marines in one of their old camps, and they were glad to find him safe.

Frémont sat up late to read his dispatches and consider his next step. Perhaps for this reason, or because they had ridden so far, no guard was kept. Before turning in, Kit fired off his gun to clean it, and did not reload, as he broke the cap tube, which rendered the weapon useless. Hardly a rifle in the camp was loaded. The men, worn out, slept in their blankets about the fires, which lighted up the small cedars all around the camp. It was the first — and last — time they failed to post a guard.

It was dark, cold, and a wind soughed through the evergreens and rippled the waters of the lake. Frémont, reading his dispatches by the firelight, thought he heard the mules stirring. He looked up — yes, there it was. All around him his men lay sleeping, worn out. Frémont did not like to rouse them. Taking a pistol, he went boldly out of camp and among the tethered animals. They grew quieter at his coming, and he could see no cause for their uneasiness. Eager to go on with his reading, he went back to the fire. Finally he turned in, never knowing how nearly he came to meeting a poisoned arrow from the Indians hidden in the cedars around his camp. The camp was quiet. The fires died down.

Dick Owens and Kit slept together on one side of their fire; Basil Lajeunesse and Denny on the other side. Beyond Denny was the fire of the four Delawares.

Suddenly Kit and Owens sat up, wide awake. To their alert ears came the sound of blows being struck somewhere across the fire. Kit called out to Lajeunesse, 'What's the matter?' There was no reply. Kit jumped up, snatched his pistol from above his saddle. At the same time, he and Dick Owens called out together, 'Injuns!' The whole party jumped up, and the Tlamaths charged them.

The Delawares were on the side of the attack. One of them, Crane, grabbed a rifle — not his own — and tried to fire at the charging Indians. But the gun was not loaded and every time he pulled the trigger, the useless weapon snapped, snapped. He swung it round his head like a club, but the Indians threw five arrows into him, and he fell, fighting bravely.

Colonel Frémont, Stepp, Godey, Kit, Dick Owens, Maxwell — were all together, and none of them needed to be told what to do. They let out a yell and charged to help the Delawares. They all fired as they ran forward.

Leading the Tlamaths was their chief. He wore a war-bonnet, carried a fine quiver full of arrows, and had an English hand-axe slung to his wrist by a cord. Kit fired his pistol as the chief rushed him, but the bullet merely cut the cord of the hand-axe. At the same moment, Maxwell shot the chief in the leg. The chief turned to run, but Stepp's long rifle cracked, and the chief fell on his face, the arrows rattling from his fancy quiver across his back.

The other Indians were firing arrows in clouds from the cedars. But when the chief was killed, they began to retreat. Before many minutes, the mountain men had loaded and taken cover behind tree-trunks around the camp. Then the Indians left them.

Denny, the half-breed, lay in his blanket, his breast full of arrows. Basil Lajeunesse had never stirred after the chief cracked his skull with the hand-axe. It was that sound which had wakened Kit. And Crane, the Delaware, had three arrows through his brave heart. The mountain men's sorrow for these three brave fellows was great, and no doubt they all blamed themselves for neglect in not posting a guard. They covered the dead men with blankets and waited for the dawn.

In the morning, after their long, anxious night, they made sure that the Indians had gone. Then they came back to their cold dead fires and the stiff figures under the blankets, and stared at each other's hard, set faces. They all looked old and tired that morning.

But Kit walked out to where the Tlamath chief lay, stiff and yellow on the blood-stained ground, his fine arrows all awry across his back, his steel hand-axe lying near his hand. Kit caught up the axe and savagely knocked the dead man's skull to pieces. Afterward, Sagundai, the Delaware chief, scalped it.

This sort of thing will be shocking to people who have never seen their best friends murdered in their beds without provocation; to people who have never wakened to find themselves facing a rain of poisoned arrows — and an empty rifle in their hands; to people who like beefsteak, but have not the nerve to kill — or watch another kill — a beef; to people who favor capital punishment, but would not spring the trap beneath the condemned man: to people, in short, who have not the courage of their convictions.

Hatred and fear are one and the same thing at bottom, and if Kit and his fellows showed savage hatred for the dead chief, mutilating his body six hours after the fight, it is more profitable for us to understand than to condemn.

For twenty years Kit Carson had lived in a region where every stranger was an enemy, where he could never cross a ridge without wondering what danger might assail him, where he never went to sleep without the possibility of being killed in his blankets. His life was one of unceasing vigilance, and all his early portraits show that intense gaze, that alertness and watchfulness — as of a wild animal — which the old-time mountain man invariably showed. Now, at the only moment when that watchfulness was relaxed, he finds his life assailed, his friends murdered. But for pure chance, it might have been himself.

Kit Carson had, at this time, killed some twenty men. Many of them he himself had scalped. He had also seen many a body mutilated by the Indians, and thoroughly understood the frightfulness which alone intimidated them. A hunter who butchered his own game, he was accustomed to blood, to the use of the knife on flesh. What more natural than that he should snatch up the very axe that had killed his friend and attack the body of his murderer?

It is not a pretty picture. But it is a little less lurid seen against its background of wakeful nights, hard riding, anxiety, fear, hatred — all these maintaining their pressure for years on end. Twenty years of it. Twenty years.

The dead chief was recognized as one of a party whom Frémont had fed out of his scanty larder but a few days before. The sign about the camp showed that the war-party had numbered about twenty braves. No doubt, after such a *coup*, the Tlamaths would harass the expedition all along the route. The Americans laid the attack to the animosity of the British.

Packing the three dead men on mules, they set out for their main camp. But the timber was so thick that it bruised the dead bodies, and so the men paused, buried them, and concealed the grave under brush and foliage. For of course the Indians would dig them up and mutilate them, if they found the grave.

Having joined the main party, Frémont pushed on, leaving an ambush of fifteen crack shots behind in his camp. After a few miles he stopped. Soon after, the ambush squad reported with the scalps of two Indians who had ventured within range of their rifles. But it would take more than two or three Tlamaths to pay for Lajeunesse, Denny, Crane.

And now Frémont, having had time to interpret the ambiguous dispatches from Washington — dispatches that (never made public) apparently left him free to raise the flag of revolution in California if he saw fit — decided to return to California and start the show. He set out, passing along the other side of Klamath Lake, so as to avoid going past the scene of the late tragedy. He made camp at a point opposite that scene.

Kit was sent forward with ten men under orders to send back word if any Indian villages were found. In case the redskins saw him, he was to act at his discretion. He had a free hand. The orders suggest that Frémont had news of a village near by, or suspected there might be one. At any rate, Kit found one of about fifty lodges within a few miles. At the moment of discovery, the agitation of the Indians in their camp showed him that they had seen his party. Accordingly, he 'concluded to charge them, done so.' The Indians made a stubborn stand, but, after losing a few men, they ran.

Their lodges were made of the long leaves of the flag; dry and nicely woven. In the lodges Kit found all manner of fishing tackle and huge quantities of fish. He immediately set fire to the lodges, which burned furiously. Kit and his comrades stood and watched the blaze with satisfaction in their hearts. Kit always said that this conflagration was one of the prettiest sights he ever saw.

Frémont saw the smoke and came hastening to join the fun, but it was all over when he got there. He moved on a few miles, posting another ambush near the burned village. Dick Owens was in charge of it, and soon sent back word that fifty Indians had returned. Kit led the men back at the gallop to complete their revenge on the Tlamaths.

On the way they met one Indian. Kit charged him, the other men close at his heels. At a distance of three yards, Kit fired. But his rifle — which had been repaired as best he could without parts or proper tools — would not go off. It snapped. The Indian's bow was drawn, ready to throw an arrow into Kit, and Kit knew that the arrow was poisoned. Like a flash, Kit flung his body behind his horse, Comanche-fashion, to avoid the shaft. Frémont, seeing Kit's danger, fired at the Indian, but — in his haste — missed. The Indian was on the point of loosing his arrow at Kit, who could not have escaped, so short was the distance.

But Frémont, though unarmed, was mounted upon his gray war-horse, Sacramento, and that horse was not afraid of any Indian. Instantly the Colonel turned the animal upon the Tlamath and rode him down. The arrow missed its mark. Kit's life was saved. And before the Tlamath could regain his feet, Sagundai, the Delaware chief, flung himself from his saddle and beat him to death with his war-club. The mountain men showed quarter to no one. Their hearts were hot against the Tlamaths.

This sort of campaigning will explain the feelings of the Indians toward their enemies. For the mountain men were much like Indians, because they lived under the same conditions. To the Indian, war was always like that: a bitter, personal, family feud, marked by age-old treachery of his foes, by the cruel and wanton murder of sleeping men, children, and beloved women, by a rankling hate no treaties could assuage. It used to be the fashion to hold up holy hands in horror at the savages. But now that our own aviators are trained to bomb non-combatants from the air, we are beginning to have some faint conception of the terror and cruelty and beastliness of primitive wars. Such warfare is savage, revengeful, vindictive: the mountain men understood it perfectly.

Next day Maxwell and Archambeau killed another redskin, who had boldly laid down his game when he saw them coming and attacked them with arrows. The first shot dropped him, and he was scalped at once.

Godey and Kit gave chase to another band, but their mules could not overtake them. One brave remained behind to fight, hiding behind a rock, and laying his arrows fanwise on the ground before him where they would be handy. Kit and Godey had to run back, dodging, to avoid his shafts. But the moment Kit had a chance, he dismounted and fired. Says Kit, 'My shot had the desired effect. He was scalped.... He was a brave Indian, deserved a better fate, but he had placed himself on the wrong path.'[1]

Next day Frémont's hunters surprised and killed another Tlamath who was watching their camp. The white men were starving, but had to come back, having been unable to find other game.

In May, Frémont was in California. In June, the Anglo-Saxons raised the Bear Flag in open rebellion. The same day, the Oregon boundary was settled by treaty with Great Britain. The Mexican War was being waged with brilliant success everywhere, and the Spanish settlers trembled for their property and persons.

Two Americans were taken by the Mexican soldiers and cruelly put to death. Though not of Frémont's party, their murder aroused great resentment. A boatload of the cussed Spaniards was seen approaching San Rafael Mission, and Kit Carson with a few others was sent to intercept them. As the men left their boat, Kit and his party rode up, dismounted, coolly took aim, and shot three of them down. One of the three was an old man.

This cold-blooded killing of non-combatants is the only blot upon Kit Carson's long record as a fighter. Of course, there was provocation. And Kit was not a soldier. All his fighting up until that time had been

[1] Grant, *Kit Carson's Own Story of His Life*, p. 75.

with men not in uniform. To Kit, an enemy was an enemy, and he could have pointed to plenty of instances of white men shot down by Mexicans under like circumstances. It has also been claimed that Frémont told Kit that he 'had no room for prisoners.'[1] Nevertheless, it was ferocious.

Frémont blames the Delawares; Senator Benton stresses the provocation; Kit omits all mention of the incident. Evidently, when his hot blood had cooled, Shame and Blame, those two guides which he had hitherto followed, made him feel that the less said about the affair the better.

But the true motive for his action on that occasion was probably the mountain man's ingrained distrust of and contempt for the Mexicans. As the saying ran, 'If Spaniards warn't made for shootin', what are rifles for?'

Soon after, Kit was one of a dozen picked men who went with Frémont to spike the guns of the Castillo of San Joachim across the straits of the Golden Gate. Rowing up to the Fort Point in the gray dawn, they scrambled up the slope in time to see the Mexicans riding off at full speed. The rat-tail files used to spike the guns had been supplied by a U.S. naval officer. Certainly, Frémont had plenty of sympathy for his stand in supporting the revolution in California.

At Sonoma, Frémont's men celebrated the Fourth of July with a big fandango, and next day was organized the California Battalion of Mounted Riflemen, with a roster of two hundred and twenty-four men. On July 7th, at ten o'clock in the morning, Commodore Sloat landed at Monterey and raised the American flag there.

[1] Richman, *California under Spain and Mexico*.

Within a week Old Glory was flying at San Francisco and at Sutter's Fort. Shortly after, the British flagship sailed into Monterey, its commander having intended to annex the whole of California for his King. Frémont, satisfied that the Sacramento Valley could take care of itself, led his men to Monterey, and joined forces with the Commodore. There Kit Carson joined the Navy!

Commodore Sloat was a little frightened over the turn things were taking, and, resigning his command to Commodore Stockton, went home in a state of bad health. But Stockton was not a man to worry. When Frémont turned up, Stockton reminded him that a commodore outranked him, and immediately enlisted Frémont's men as members of a brand-new and quite unique command — the Navy Battalion of Mounted Riflemen!

Worse than that, he marched Kit and his comrades upon the deck of the sloop Cyane — a vessel which the landlubber Kit describes as a 'frigate' — and sailed away with them to San Diego, a distance of four hundred miles and more. The wind was fresh, the sea rough, and few of the mountain men had ever seen salt water. That voyage nearly finished Kit, who was sick as a horse all the way. He told Maxwell that he would never set foot on a ship again — 'not as long as mules have backs.' Kit reckoned that, if he could once git back on the prairie, he would ruther fight all the Kioways alone than go to sea. So long as he lived, he never would board another vessel. His career in the Navy was brief, unhappy, not to say ignominious. And when he got to San Diego there was no fighting to be done, after all. The Mexicans evacuated the place.

The whole of California was now publicly declared a

territory of the United States, and at once it was decided to send Kit to Washington with dispatches which would officially announce the conquest.

This honor came to Kit as a tribute to his courage, address, and competence. It was no empty honor, for it meant a dangerous ride across the continent at record speed. Kit was made a lieutenant to give dignity to his errand. He was put in command of fifteen picked men, including half a dozen Delawares, and allowed *carte blanche* for expenses.

And the journey was not merely an honor, but an adventure as well. Kit was to present his dispatches directly to the President himself. He was sure to be the hero of the hour, enjoying a public fame which no mountain man had ever experienced. And on the way he would have a chance to spend a few hours with Josefa at Taos.

On September 15, 1846, Kit and his band rode out of Los Angeles, bound for Washington, bound for the White House, conquering heroes all.

That day Kit Carson, behind his quiet manner, felt sure that he could shine in the biggest kind of crowd!

V
SOLDIER

CHAPTER XIX
SAN PASQUAL

Not far from Socorro, on the Rio Grande del Norte, Kit Carson met the Army of the West advancing toward the coast, its object being the conquest of California! General Stephen Watts Kearny, its commander, was somewhat chagrined to learn that Frémont and the Commodore had already accomplished his mission. The West-Pointers did not love Frémont, who — they thought — had had more than his share of public honor and opportunity, not to mention the political backing of Senator Benton. Kearny's dislike for Frémont probably extended to his right-hand man, Kit Carson. At any rate, he immediately ordered Kit to face about and guide his command to the coast. His reason — or excuse — was that Kit had just passed over the country which the Army of the West was about to traverse.

Kit was new to military life and naïvely informed the General that he could not turn back, having pledged his word to carry the dispatches to the President in person. But Kearny, assuming a hauteur which might have impressed a lesser man than Kit, told him that he, General Kearny, would assume all responsibility for the matter. Then the General ordered White Head Fitzpatrick, his guide so far, to carry Kit's mail to Washington.

Kit said nothing. Kearny's conduct shocked him profoundly. Kit could not understand a world in which one man could assume the responsibilities of another so lightly. He had made a solemn promise. And here came a suggestion that he need not keep it! And from a General, too! It only confirmed Kit's low opinion of military men. On the coast he had come to think the men of the Navy and Marines bold, noble fellows. But this Kearny! Kit considered that the man was merely taking selfish advantage of his rank.

That night Kit made all preparations to slip away and go to Washington regardless: he knew that none of Kearny's men could catch him. The idea of obeying an arbitrary order against his better judgment seemed sheer nonsense, especially as he had no respect for the General. But after talking it over with his men, he decided that he *ought* to obey. After all, he knew the country better than White Head.

White Head, who seldom laughed, grinned with satisfaction on escaping from the Army of the West, and set out next day for Washington. Kit gave one envious look at the old coon's back, then faced about and put out of his mind all thoughts of Josefa, the President, honor, and novel scenes.

The Army of the West moved on. Kit laughed at Kearny's outfit. Wagons, cannon, foot-soldiers: the usual ridiculous equipment of the Army in the West. To the General he said, 'It'll take ye four months to git to Californy at the rate you're goin'.'

Kearny made a lightning calculation: two plus two are four. Men have two feet; mules have four. Obviously, men on mules would be quicker than men afoot! After that, the General used pack-mules instead of wagons, and the column made progress. Kearny had marched only one thousand miles or so before he thought of that. He was quick-witted: that was why he was a General. He — and of course Kit — deserve all the credit for this improvement in the Service; it was an improvement which the War Department did not whole-heartedly endorse until after the Civil War. Afterward, Kit commented upon this affair to Sol Silver: 'Whenever thar's war, Sol, the civilians take holt and fix things up; if it warn't for the civilians, the Army men would still be fightin' with bows and arrers!'

Kearny was disappointed when he heard that California had been pacified. But late in November he captured Mexican dispatches which told him that it had all to be done over again. On first hearing of the conquest, he had reduced his Army of the West to little more than a hundred men: now he had leisure to reflect upon his folly. But Kearny pushed on. He was as rash as he was unprepared. And because he had taken possession of New Mexico without a battle, he supposed that he could conquer the fiery Spanish Californians in the same manner. But even in New Mexico, a revolt was brewing, and when he got to California he was soon undeceived.

At the little Indian town of San Pasqual, about fifty miles from San Diego, Kearny's advance guard established contact with the pickets of the enemy. Captain Johnston was in command, with fifteen men, of whom Kit was one. As soon as the General learned that the Mexicans were just ahead, he and his staff joined the advance guard and he ordered a charge. The Spaniards were running away. He could see them.

Left to himself, Kit would never have walked into

such an obvious trap. But Kearny did not consult Kit. Kearny was lusting for a fight. All the other generals had been shooting down Mexicans by hundreds; he had done nothing but march and read proclamations. Now he would show Frémont how to take California! 'Charge!' And away dashed the score of Americans, mounted upon mules, in the wake of the spirited Spanish horses. Twenty — against eight times their number!

The ambush worked perfectly. The Americans dashed into the village headlong, and the enemy withdrew, galloping easily ahead of them, leading Kearny on to his destruction. Happily for Kit, his horse stumbled and threw him. His rifle flew from his grasp. He struck the road in the path of the charging troopers. For of course Kit was in the lead.

About his head he saw the madly flying hooves of the charging animals. He thought they would kill him, sure. But rolling to one side of the road, he avoided them. They swept past, in a cloud of dust. Kit got up, found himself unhurt, looked for his rifle. It was smashed — the stock and barrel lying in two places. He ran on after the others, who were now a hundred yards ahead and still racing. Finding a dead dragoon on his way, he grabbed up the man's cartridge-box and carbine, and hurried on.

Captain Moore, with reënforcements, swept past him, smothering him in the dust. And far up the road he could hear the shooting, see the Spaniards and Americans racing, the latter all strung out along the road. Then, suddenly, he saw the Spaniards wheel, come back at the gallop, their long lances tossing athwart the sky, above the dust of their coming; saw their lances lowered and piercing the helpless dragoons, who swung their carbines uselessly against the sharp blades at the ends of the long staves; saw Kearny and the other officers using pistol and saber in vain to defend themselves! Why, the cussed Spaniards were rubbing them out like flies, killing them like buffalo! Kit had never seen such a slaughter. He made all the haste he could to join the spree.

By the time he came up, the road was strewn with dead and dying. Of forty Americans engaged, thirty-six were dead or badly wounded. It was the most terrible casualty list Kit had ever seen among white men. From that hour the Army of the West divided its time between the duties of a burial detail and those of a hospital corps. Kearny himself had been wounded in two places. He had found the fight he was looking for. But Kit — and old Antoine Roubidoux, who had had the bad judgment to sign up as Kearny's interpreter — shook their heads.

All that saved the command from annihilation was the appearance of the howitzers, which galloped up furiously. The Spaniards rode away when they saw the cannon coming. But they need not have been alarmed. Before Davidson could unlimber his guns, one of the teams ran away into the midst of the enemy, and all his gunners were lanced or shot. By good luck the Spaniards had no gunners who could use that half of the artillery they had captured.

Kit and the others took cover among some rocks near by and remained there all night, not daring to move. They spent the night burying the dead. Next day Kearny marched on. Kit was now in command of the advance guard of fifteen men. Johnston was dead.

Before the command could reach the water on which they intended to camp, the Mexican cavalry charged them and drove them back to another hill. There, among the rocks, they passed another night, waterless, with only mule-meat to eat. The men they had sent to the Commodore had been captured within sight of their hill.

Kearny, rash as ever, decided to march on, be the consequences what they would. He had learned nothing from his previous experience. But his officers had, and, backed by Kit and Antoine, they virtually forced him to remain. Their one hope lay in sending word to San Diego, in getting succor from the Commodore. Lieutenant Beale volunteered to go. He had an Indian boy who would go with him.

Beale was a naval officer. But Kearny had come to realize at last that a man like Kit Carson, who knew something about the country and the ways of Western fighting, might be of some value on such a mission. And so, when Kit volunteered to go, Kearny approved. Godey and others had failed; maybe Kit could make it.

As soon as dusk fell, the three men crept out of camp and down the hill. All around the hill they could see the Mexican sentries, three separate cordons of mounted lancers, covering a distance of two or three miles. For of course their commander knew very well that Kearny must get help or surrender very soon. Kearny had no water, no food; half his men were hurt or sick.

Those lances swinging against the sky were full of menace. Those same sharp blades had stabbed the life out of half of Kit's comrades. Kit knew that if he or Beale or the Indian were seen, it would be the end of them — and the end of the Army of the West. At the first alarm the lancers would gather like hornets, would kill the three for spies. They had no chance at all if once they were discovered. They must get through.

Their way lay over rocks and brush. Their shoes made noises which seemed very loud to them. Kit whispered to Beale, and they removed their shoes, tucking them under their belts. Slowly they crept down the hill, through low bushes, among sharp stones, picking up cactus spines as they advanced.

In approaching the first line of sentries, they were very cautious. Their canteens rattled against the rocks, and they unbuckled the straps, leaving the precious water behind. After what seemed hours, they drew near the second line of lancers. These were more active, as they had more ground to cover, and kept riding up and down, often passing within a few yards of the party, who flattened themselves against the rough ground. The Spanish commander himself rode past, and they heard him warn his men to scour every foot of the hillside. He knew that Kit would be the man to go. 'Be watchful,' he urged them. 'The wolf will get away!' His men rode everywhere, and Beale gave himself up for lost.

Still Kit led on, worming forward on his belly. His knees and elbows, his bare feet, were filled with cactus spines, which he could not avoid in the darkness. Stones cut his flesh, thorns scratched him. But he kept on, inch by inch, until the second cordon had been passed. Beale followed, and the Indian boy came last, thankful that the others had gathered most of the souvenirs along the way.

The third cordon of sentries lay just ahead. As they

crawled through it, one of the lancers almost rode over them. Kit held his breath. He could hear Beale's heart beating in the silence, and it seemed to him then that it missed a beat. Only a miracle could prevent the horse from taking alarm at the three prone figures. Yet it did not snort. Kit waited. But the man did not ride on. He reined up his animal, dismounted, and got out the metal-covered wick he carried. Beale and Kit saw that he was going to strike a light!

The Spaniard was a strapping fellow, and in no hurry. Striking steel on flint, he ignited the wick, and prepared to light his cigarette. But first he glanced about him, holding the flaming wick in hand. To Kit and Beale it blazed like a searchlight. It seemed impossible that they were not seen.

Suddenly, the sentry quenched the wick, mounted, and galloped over to his nearest comrade. They could see the two lances against the stars. Had he seen them? Was he quietly summoning his friends? Beale whispered to Kit, 'They've got us. Come on. Let's jump and fight.'

But Kit was cool. 'No,' he whispered. 'I've been in worse fixes. The boys back thar are countin' on us.'

The two horsemen talked for a time, then separated, and rode off again about their business. Kit and Beale breathed once more.

Within half a mile they could see the trees along the stream. There they would find shelter, water. There they could stand up, stretch their legs, pull some of the spines from their tortured flesh, make trail for San Diego. From there all would be plain sailing.

At last they reached the trees, took cover, began to

talk and laugh. Then, at the same moment, both found that their shoes were gone, had slipped from under the belts! They were barefoot — and it was some thirty miles to San Diego. Thirty miles — over rocks, and gravel, and cactus beds. Thirty miles of such walking — barefoot — in the dark!

They set out.

Progress was slow, and they had no food. By dawn, every step was torment. In the afternoon, the inflammation of their feet made them feverish. By nightfall, Beale was out of his head for minutes at a time. To him, the whole world seemed paved with prickly pear.

After nightfall they saw the lights of San Diego shining far ahead. Here they must renew their caution. For they were in no condition to fight, and as for running from an enemy — that was simply impossible, their feet were so swollen. Kit advised Beale that each one must go in alone, and each one from a different direction.

Beale was hurt most, and they started him straight for the lights. Kit turned to the left, and the Indian to the right. Then they parted, making for the little town where — they hoped — the Commodore was still in command.

The Indian got in first, for he had retained his moccasins and therefore was less exhausted than the others, although a moccasin is poor protection against cactus in the dark. Soon after, the pickets challenged Beale, who was carried in, almost unconscious, utterly used up. How he got through the Spanish sentries posted around San Diego is still a mystery. It was nearly morning when Kit got in. He had taken the longest way.

Kit Carson was laid up for some time, but Beale was

a mere shadow of a man for a year and more. All three had shown the stuff of which heroes are made. In the heat of conflict, physical courage is cheap. But to walk for thirty hours (eighteen of them in darkness) through cactus and desert rocks, helping and encouraging a weaker companion, without food, without water most of the time, barefoot, and to keep on after exhaustion has sapped one's strength — that is heroism, that is manhood. In his memoirs Kit gives this adventure nine modest lines.

But the world got news of it, and this second story made Kit a famous man. Before, Frémont had advertised him on equal terms with Godey. But this time Kit had succeeded where Godey failed. From that day, Kit's fame surpassed that of all the mountain men.

Kit, however, was not concerned about such details. Kearny was rescued and came on to San Diego. After a month, Kearny joined forces with the Commodore. The Commodore, directing the campaign, chased the Mexican forces out of Los Angeles, and they surrendered to Frémont, whom they liked better than Kearny. Frémont brought his prisoners in to Los Angeles, and immediately Kit left Kearny and joined his old commander.

Kit had had his bellyful of Kearny, who had bungled everything he tried to do. He pacified New Mexico by reading notices and shaking hands, and then rode away, leaving the country practically unguarded. He would still have been marching for the coast if Kit had not reorganized his transport. And he had led his command — already cut in two by his folly — into a trap from which he was powerless to rescue the remnant. As commander of Kearny's advance guard after that *débâcle,* Kit had fought all his battles that were at all successful, and then pulled him out of the hole at the end of it. He was fed up with Kearny. To look at the man made Kit feel as sick as a cow with a Galena pill in her lights. Yet there are people who think that Kearny was a soldier, and that he conquered California. Even the Mexicans would not acknowledge defeat to such a fellow. They went out of their way to surrender to Frémont, his subordinate.

Kearny set up as Governor of California, but Stockton named Frémont Governor, and Kearny had to yield. However, Kearny's day was coming. Though he could not fight, he *could* make life wretched for his subordinates. When Stockton sailed away, Kearny got control of things, and made Frémont and his 'topographical party' march at the tail of his own column all the way back to Missouri. At Fort Leavenworth, Frémont was put under arrest. West Point would have its revenge. It was unethical for a topographical engineer to conquer California, especially when regular line officers could not.

In the meantime, Taos was ablaze with revolution. Kearny's pacification was of paper, and stimulated by Martínez, the *padre* who had married Kit, the Pueblo Indians rose one night and murdered the Governor, Charles Bent, and most of the white men in and about Taos. Bent would not defend himself when the mob invaded his house. He knew the Indians too well: if he angered them by resistance, they would kill his family also. In that family group was Josefa, and after the mob had departed, she and her sister, Mrs. Bent, had to remain for more than twenty-four hours in the room with the dead, scalped body of the Governor.

After that, by the kindness of friends, the two women disguised themselves as Indian squaws and worked as servants in a kitchen until the trouble was over. This was in January, 1847. The month following, St. Vrain and a mixed army of volunteers and regulars besieged the Indians in their pueblo, took the mission church, their fort, by storm, and destroyed almost the whole male population. Afterward the prisoners were tried summarily — for Treason! — and hanged in batches. But of course the men higher up, who had planned and instigated the revolt, were never molested. Kit never ceased regretting his absence from that fight.

Kit was ordered to Washington again with dispatches, and this time Kearny was powerless to interfere. Beale, the invalid, went along, so weak that Kit had to lift him off his horse whenever the party halted. On the Gila there was the usual unimportant brush with Indians. In St. Louis, Senator Benton made Kit welcome, and at Washington, President Polk gave him his commission as Lieutenant of Rifles, U.S.A. Then Kit started west again, his saddle-bags full of mail. On the frontier he was given an escort of fifty men, and after passing Pawnee Rock was the means of saving a great part of the oxen belonging to a Government train. Two of his own horses were lost in this scrape, owing to the fact that his men dropped their lariats while firing at the Indians stampeding the oxen. Such carelessness annoyed Kit's methodical mind. At Santa Fe he gladly left his military escort of fifty men, hired sixteen of his own band, and arrived — after the usual brush with Indians — at Los Angeles in October.

In the spring he went to Washington again. On Grand River, some of the saddles and uniforms were lost in crossing on a raft, and the naked men had to ride themselves raw on bareback mules all the way to Taos. And at Santa Fe, Kit was informed that the Senate had refused to confirm his commission as lieutenant. He had no rank, and no pay coming for his two years of service. Kearny and West Point again![1] Kit thought that was pretty rotten. Politics and diplomacy were not his forte, and he had only contempt for the effeminate tricks of men who would strike at an enemy through innocent and helpless subordinates. His scorn of the army grew and grew. Hell's full of soldiers!

But the question was — what about the dispatches?

His friends advised him to turn them over to the nearest officer. He was no soldier, they said, to go carrying mail for men who would not recognize his services. He could not collect a penny for his labor. Why should he ride the long trail, through danger and hardship, for men who struck him in the back? Men whose lives he had saved?

Moreover, they assured him that the Comanches were raiding the Santa Fe Trail, so that he would have to ride far north to the Platte before turning east. Then they paused to hear Kit's answer.

'Wal, boys, maybe I ain't a lieutenant after all. Maybe I ain't even a soldier, thank God! But I reckon a good mountain man is a hell of a sight better than either of 'em. I never asked to be made a lieutenant, and when the war was over, I'd ha' had to re-sign.

---

[1] In fairness to the partisans of General Kearny, some of whom have objected to the estimate of the man here attributed to Kit Carson, I wish to call attention to the paper by Mr. Thomas Kearny in the *New Mexico Historical Review*, Vol. V., No. 1, January, 1930, page 1 ff. It presents the contrary view on this controversial subject in a most able manner.— S.V.

Them old women to Washington caint make me no better than I am, nohow. What they think don't matter a damn. It's the country I'm thinking of. I've always done my duty the best I could. I told Frémont I'd carry the mails through. I reckon I'll do it. *Adios!*"

VI
RANCHER

CHAPTER XX
THE LAST TRAP

On his return from the East, Kit Carson decided once more to settle down and farm. It was now almost six years since he had married Josefa. During all that time he had visited his home only four times, and in all those visits — including his honeymoon — had spent rather less than sixteen months with her. She and Kit loved each other, and both of them were a little tired of the endless wanderings and multiplied dangers to which his service with the army brought him. Kit located his ranch on the Rayado, now known as Rayado Creek, in northeastern New Mexico.

Kit's attempts at ranching in New Mexico long antedated the days of the range cattle industry, for the invention of which the Texans claim all the credit. Even William Bent, as soon as the travel on the plains made the raising of beef cattle profitable, wintered Mexican steers on the Arkansas, hoping to sell them to the army in the Mexican War. Kit's venture on the Cimarron was somewhat earlier than Bent's, and on a small scale. And until travel on the plains began to be heavy, there was no profit in beef cattle. The true frontiersman could live on game, and thought beef insipid stuff. But neither Kit nor Bent nor the Texans invented the cattle industry in the Old West. They simply adopted or expanded an industry which had flourished time out of mind on the great Spanish grants to the south of them in Mexico.

Kit had some cattle, but his chief interest lay in horses and mules. The Bents had made mules the favorite animals on the prairies, and this trend was further influenced by the habit of the Mexicans, who preferred mules in wild country. As the Spanish or Mexican mules were dwarfed, inferior creatures, the Bents established large mule farms in Missouri on which to raise animals of finer and larger stock. This enterprise has grown until the Missouri mule is famous the world over. Kit therefore saw no reason why he should not profit by breeding superior animals on the Western plains. Cattle, mules, horses: these were Kit's chief concerns. Maxwell was Kit's partner in the business.

It is amusing to see Kit, every so often, pausing to reflect that he must give up his roving life and settle down. The reason he usually gave was that he was getting old. He thought he was getting old at twenty-nine — at thirty-two — at thirty-three — and now at thirty-nine. Each time he was quite serious about it: he had never been so old before! But now, in 1848, Josefa was expectant, and the coming child offered an argument not readily to be set aside. Adeline, in Missouri, was now twelve years of age. Kit felt that he had responsibilities he could not evade.

By this time many of his old friends had gone under, and the West was already changing so rapidly that an old mountain man felt 'queer' there. No doubt, at heart, Kit loved the roving, dangerous life to which he had been trained, but he and his fathers were farm folk by tradition, too, and, like the Iowa farmers of to-day, Kit constantly intended to retire and settle in California! However, business, family ties, friendships, habits, all conspired to hold him to Taos and New Mexico and the mountains and plains in which he had been made a man.

The ranch was located upon the vast estate of Maxwell's father-in-law, Charles Beaubien, who had been presented by his friend, Governor Armijo, with almost two million acres, an estate as large as any one of several Eastern States, and by far the most extensive private holding ever known in North America. After his early days as hunter, trader, guide, and trapper, Maxwell lived like a lord in his palace on the Little Cimarron, the ruins of which are still a landmark to the motorist on the trail to Taos. Open-handed as a true mountain man always was, Maxwell entertained his old comrades freely for as long as they chose to stay. For once, a trapper had more than he could throw away, even with the biggest kind of heart. Wealth meant nothing to Maxwell when an old comrade of the beaver trail was concerned, and he was glad to get Kit to throw in with him.

In the spring of '49, Josefa, now twenty-one years old, gave birth to her first child, Charles Carson. Properly enough Charles, when we remember how Charles Bent had befriended Kit at Franklin, at the end of the beaver days, and countless times after, when the men were brothers by marriage.

Charles was Kit's first male child of whom we have any knowledge. But Josefa did not keep the baby long. He died within the year.

The monotony of ranching was broken during these months by a number of adventures, several of which are among the most celebrated exploits of Kit Carson.

Whenever there was any dirty work to be done, Kit was called upon to help — or, if others were sent, the senders commonly had cause to regret their unwisdom in not calling for Kit.

To begin with, he was called out to act as guide for Colonel Beall, who was chasing Apaches. The Colonel's guides reported that he could not take his command through to reach the enemy. Colonel Beall refused to believe that, called upon Kit, and set out. The First Dragoons could do anything, the Colonel insisted. Kit led him through, found the Apache village, and made the desired treaty. This made the Colonel feel cocky.

Then the Colonel marched up to the Arkansas, where was a huge encampment of Indians. The treaty between Mexico and the United States provided for the return to Mexico of all the Mexicans captured by the Indians and carried into the territories of the United States. Now the Plains Indians had waged war against Mexicans for centuries, and saw no reason why the United States should interfere with their business of capturing slaves and horses. And when Colonel Beall arrived on the Arkansas, it was clear that the Indians had no intention of giving up their plunder. Indeed, many of the captured peons preferred living in the Indian camps. If they did not, they dared not admit it in the presence of the red men. The only way in which these captives could be liberated was by purchase.

Two thousand Indians were in the camp, and the Colonel had only two companies of dragoons. But he was bold — or rash — as Kearny himself. Kit had to spend the day dissuading him from attempting to take the captives by direct attack. The Indian agent, then in the camp, warned him of the folly of his project, and all his own officers supported Kit and the agent. Finally the Colonel marched away to fight again some other day. He was wiser than most of the commanders on the prairies. He was willing to listen to expert advice.

In October, 1849, a trader named White, traveling westward with a train from Missouri, left the caravan near Point of Rocks and set out for Santa Fe, then only about one hundred and fifty miles ahead. He drove his own carriage, in which were his wife and little girl. With him rode another American, a German named Lawberger, a Mexican, and a negro servant.

When in camp on the trail, a number of Apaches came up and demanded the usual gifts. White refused them, and drove them away from his camp. Later, the same party returned with reënforcements, and repeated their demands. Again White refused to make presents. Immediately, the Apaches attacked.

They shot down the negro and the Mexican at once. The others tried to escape, but the Indians killed them all except Mrs. White and her little daughter. Afterward, the marauders laid the dead bodies in a row beside the Trail, and concealed themselves near by.

Presently a party of Mexicans came along, and, finding the dead bodies and the camp, began to plunder the deserted carriage. Then the Apaches fired arrows at them. The Mexicans ran away, leaving behind a boy who had been hit by an Apache arrow. The boy shammed death: owing to the fact that the Indians did not scalp or mutilate the dead, he survived. The fact that the Indians did not mutilate the bodies showed how easy their victory had been. For as a rule mutilation was inspired by spite after losses, or animated by a superstitious fear that a great fighter — one 'hard to kill' — might come alive again. The White party had offered no resistance.

The wounded boy, with the arrow sticking between the bones of his arm, came up with some Americans, who took him into Santa Fe safely. The word of the tragedy was carried to Major Grier, of the First Dragoons, then in Taos. Immediately, Grier organized a pursuit of the Indians, taking along Leroux and Fisher as guides and trailers. Leroux was chief of scouts. On the Rayado they found Kit Carson, who joined them gladly.

Having reached the scene of the murders, they picked up the trail and followed it for twelve days, clear to the Canadian River. The weather was bitterly cold, and there was some snow on the ground. The Indians had had a week's start of the column, and they did all they could to make following their trail difficult. Moreover, it had snowed since the Indians passed, though the snow was light.

Of all Indians, the Apaches, Kit always said, made the most difficulties for their trackers; and of all the Apache trails that ever he saw, this one was the worst, the hardest to make out. The trail was a week old to begin with, and here and there was obliterated by drifted snow, by the tracks of passing buffalo. But the devices of the Indians themselves made more trouble for the scouts than everything else.

Of course, the greater the number of men and animals making the trail, the plainer was the sign. The Indians, in order to confuse their trailers, made a practice of dividing up into small parties, agreeing to meet again at a point which might be many miles away. Thus, one by one, or in small parties, they would swerve from the line of march, sometimes leaving the scouts who followed in doubt as to which trail was that of the main body. If Kit and his comrades surmounted this difficulty and stuck to the main trail, they were sure to find, sooner or later, that the Indians had scattered like quail in all directions, so that even the main trail split up into as many radiating trails as there were Indians.

Then the scouts would have to guess and confer, and decide which was the likeliest direction the main party would take. Having chosen this, they would have the problem of tracing the line of march of a single warrior, who was sure to cover his trail as best he could, walking or riding on the hardest ground, through streams, in front of herds of bison or other animals, seeking in every way to obliterate the slight traces which he left behind him.

If this could be followed through, the scouts would find the trails converging again. Then, next day, the same process would have to be gone through with once more. Twelve days of this sort of thing gave the three men plenty of chance to show — and compare — their skill.

Leroux was proud of his place as chief of scouts, and a little jealous, perhaps, of Kit Carson's fame. Kit, on the other hand, had been lionized repeatedly in the cities of the East, and he was a little bit vain of his rating as the best mountain man in the West. He had had more experience than Leroux, and even Fisher, one of Bent's best men, could not compare with Kit as a trailer.

Kit Carson was a volunteer on this expedition, and felt that he was free to express his opinions, even though they might vary from those of the chief of scouts. No one has ever suggested that Kit was disloyal to any man whom he thought worthy to lead him. But his opinion of Leroux was not very high. Perhaps Leroux felt this. At any rate, there seems to have been rivalry between them, felt — if not expressed.

By the time the Apache trail grew 'warm,' Mrs. White had been in the hands of the Indians three weeks. In every camp they found pieces of her clothing, and supposed she had left these in the hope that they would know she was alive and would not abandon the pursuit. Kit and Fisher, who had lived among the Indians, could faintly guess what the poor woman had been enduring all that time: starvation, fatigue, cold, cruelty, lust. They pushed on as fast as they could.

On the twelfth day Kit sensed a change in the air. A norther was coming, a blizzard which would obliterate the trail altogether, and would force the dragoons under cover. That storm would save the Apaches unless Grier and his men could strike them first. The Major made all the speed possible.

At last, Kit and his party saw the Indian village. Kit was in the lead. As he had been the first to find the scene of the murders, so now he was the first to make out the hostile camp. Kit immediately put spurs to his horse, calling to the men to follow. He knew the effect of surprise upon the Indian. He knew that peace must follow victory.

But Leroux opposed Kit's plan and the Major hesitated. Immediately the Indians saw the troopers halt, they began to pack up and run away. Kit could not capture the camp alone. And before the Major could make up his mind to charge, a bullet from an Indian rifle struck him. The ball was almost spent, and, passing through the officer's clothing, merely bruised him, knocking the breath out of his body and making him sick at the stomach. By the time he recovered sufficiently to order a charge, the Indians were safely on their way. One Indian was killed. The body of Mrs. White, still warm, was found not far off, with an arrow through the heart. The Indians, fearing her recapture, had killed her. She was found some distance beyond the camp, and Kit believed that, had the charge been made at once, the Indians would have been compelled to leave her among their tents, alive.

No one questions the good intentions or the gallant spirit of the Major. He was simply ignorant of Western Indians and prairie warfare. He hesitated, and Mrs. White was lost. Kit did not blame Leroux, either. In fact, he defended them both: he said they did not know enough to take his advice!

The pursuit was kept up for some miles. Then the storm intervened, and the Indians got away. Kit and the others returned, fired the camp, and destroyed the animals and food and clothing left behind. They knew that the blizzard would account for some, at least, of the naked Indians. Mrs. White was found to be horribly wasted, diseased, and so brutally handled that she could hardly have lived much longer, even if rescued.

In the camp a *book* was found! This book was a novel of the Wild West, and represented Kit Carson as a great hero, who had slain his thousands of redskin hostiles. Kit looked at it with curiosity, and wondered how it got there. It was the first time he had ever encountered himself in print, and he enjoyed the novel sensation as the Major read to him. His naïve delight in this experience is shown in his words which Peters quotes in regard to it: 'Perhaps Mrs. White, to whom it belonged, knowing he lived not very far off, had prayed to have him make his appearance and assist in freeing her. He wished that it might have been so.'

Did Kit Carson, far down in his heart, remember some old folk tale brought to North Carolina or Kentucky and recounted by his mother, and picture himself as a knight rescuing damsels in distress? Was it some old ballad which stirred his childish imagination and drove him on to deeds like this? Ah, that last infirmity! 'He wished that it might have been so!'

The blizzard killed others than Indians. Grier's party wandered blinded in the icy sleet, and one man froze to death. It was the worst storm Kit ever weathered, but by good luck the command reached shelter and Las Vegas. Then Kit went back to his mud-walled *patio* on the Rayado.

During the same winter, Indians ran off all Kit's best horses, and with a detachment of dragoons, Kit overtook the thieves and recaptured the stock, killing five Indians in retaliation for the wounding of his own herders. The same winter Kit had cause to mourn the loss of one of his earliest friends, Bill New. He was rubbed out by Indians on the Rayado.

Kit lost another of his retainers, Tim Goodale, in the summer of 1850. Old Bill Williams had gone off into the mountains alone, and had been found with a Ute bullet through his body, sitting bolt upright against a tree in his favorite mountain park, frozen stiff and half covered with snow. One by one Kit's men were dying or drifting away. Bent's Fort no longer could maintain a large staff, and running meat was no longer a business worth the attention of Kit's band. The good old days were passing swiftly.

Goodale and Kit drove mules north to Fort Laramie on the Platte. Fort Laramie was the principal station on the Oregon Trail, and the emigrants who had come so far were generally in need of new stock. Kit and Goodale drove the five hundred miles without mishap, disposed of their mules at a profit, and were ready to return. But Tim Goodale, hearing so much talk of California and the gold rush, decided to go to the coast. Kit would have gone too, if he had not had a family and a ranch on his hands. He had been foot-loose so long, that now a little property seemed absolutely to fetter him. Kit rode back to New Mexico. The way home was very dangerous, as the Apaches were out, and all travel was given up for the time. Kit, however, with one man, managed to get through, traveling by night and spending his days in a tree-top on the lookout for his enemies. Of course, when he got home, he found that the cussed Injuns had run off all his stock again!

The soldiers had made no effort to recover the animals, as the redskins were out in force. Kit got Major Grier to act, however, and by hard riding overtook the thieves, and recaptured all his animals except those that the Indians had eaten. Such expeditions were generally a loss to all concerned. The Indians lost their hair and their stolen horses; the white men broke down two mounts for every one they recovered, and those recovered — often enough — were so used up that they were worthless.

It is interesting to find that Kit Carson, in addition to being among the first white men to follow the trade of the cowboy, was also the first man of that long line of heroic marshals who brought law and order to the lawless plains.

The great wave of migration to the coast, which followed Frémont's expeditions, brought an immediate change to the Southwest. The country was now under the control of the officers of the United States, and Americans no longer had to fear the indignities and extortions of the Mexican officials. Immediately, the character of the population began to alter. Fugitives from justice, drifters, the unfit and unscrupulous, scalawags and scoundrels, began to come in. All the old decencies of the heroic age began to be assailed by those who neither knew nor cared why they had been maintained. Game was recklessly slaughtered and left to rot on the ground; crimes were committed and laid at the doors of the Indians; class hatreds, race feuds flourished. Security bred wealth, and wealth tempted thieves. The age of heroes was followed by the age of banditry. It was the aftermath of the Mexican War, the most shameless war of aggression in which the United States ever engaged.

Two rich merchants set out from Santa Fe for the States, and the story went that their packs were heavy with Spanish silver, perhaps with gold from California! There was much gabble about this in Santa Fe, and one Fox, a man of desperate character and no conscience, laid plans to rob them as soon as they were beyond the reach of the officers in the Mexican settlements.

The Last Trap

In order to carry out his plans, he offered his services as guard and guide to the unsuspecting merchants, who — like many other people from the States — could not distinguish an honest mountaineer from a bad man. They hired him, and he set about enlisting a large party of desperadoes to help him rob his employers. The fact that he got together some fifty followers shows how the character of the population was changing, and also how cowardly the scoundrels were.

In Taos Fox tried to interest a man he knew, but for some reason the fellow was cautious and would not join. When Fox and the merchants had gone far enough to be beyond the reach of aid, the man told the story in his cups, and it came to the ears of Lieutenant Taylor, of the First Dragoons. At once the Lieutenant spoke to Kit, saying that he wished Kit to go and arrest Fox for debt. Kit refused.

Then the Lieutenant, caught in his own trap, made a clean breast of the matter. Kit investigated, and arranged to go.

The plan was that Fox would lead his men to the banks of the Cimarron and murder them there. Then he would ride away to Texas with his booty. Already the caravan must be nearing the river. Kit took ten dragoons and rode day and night on the trail of the murderers. On the way he met one Captain McEwell with a detachment of recruits. The Captain joined Kit, adding twenty-five men to his posse.

On sighting the camp, Kit rode in alone, allowing his men to follow quietly. In the camp he found Weatherhead and Brevoort, the intended victims, and asked to see Fox. Fox was pointed out. By this time the soldiers were dropping into camp. But before Fox could see what that meant, Kit covered him with his pistol and put him under arrest. Fox grumbled, but said nothing. He saw the game was up.

The two merchants were astonished, alarmed, and then deeply affected by what had happened. They pressed Kit's hands, and said he had saved their lives. Weatherhead, whose unpublished account is in my possession, was the first to speak. With tears running down his cheeks, he asked Kit how he could reward him for his services.

Kit was embarrassed, and, looking off across the prairie, he said, 'Gentlemen, you don't owe me anything. I only done my duty. As for a reward, if thar's a God in heaven, He will reward me hereafter.' The men stood around, and many of them shed tears at Kit's simple faith.

But Kit, practical as ever, advised the two merchants to choose from their escort a few men whom they could trust and send the others about their business. This they did, and some thirty rascals got their marching orders. Captain McEwell took Fox off with him, and Kit rode home to Taos and Josefa.

This adventure and the recognition it brought him gave Kit a good deal of satisfaction. It was typical of his character, simple, sincere, direct. With Fox, might made right: with Kit, right made fight. Kit's strength was as the strength of ten, because his heart was sure.

Nowadays many people complain of the want of civilization in our Nordic character and culture. They rail at the American lack of ideas, the superstition and credulity which, they say, mark our national mind. There is much to be said for this point of view. But it is worth asking what value ideas have in the conquest of a continent — or a world, for that matter? The strength of the Nordic peoples has always been the fighter's strength, and ideas are of very doubtful service to a fighter — or a woman in childbed. Nobody will deny that the French have plenty of ideas, culture, civilization. The French occupied North America first. But to-day the Indians — who had no ideas at all — are more important factors on the continent than the French. It is a ten to one bet that the same thing will prove true of Africa, South America, Asia. Ideas — until they have crystallized into beliefs (and by that time they are usually questionable) — simply will not shoot plumb-center. The critics of Nordic credulity are the enemies of Nordic success.

Kit was so superstitious that he was afraid of the giants said to live on the island in the Great Salt Lake, so superstitious that, if he missed a fair shot in battle, he would not fire at the Indian again, but let him go unscathed. All his ideas were childlike, traditional, unquestioned beliefs. He believed in God, in the right, in courage, honor, integrity, the sanctity of pure women. He believed in justice and mercy, and thought he knew them when he saw them. Yet, somehow, circumstances always made monkeys of the men of culture and ideas who failed to take his advice.

That is to say, Kit could deal with everything real: the men of ideas could cope only with the non-existent giants! His was the spirit which makes nations great — and individuals tiresome.

In the spring, Weatherhead and Brevoort presented Kit with a pair of handsome silver-mounted pistols. Shortly after, he took a train of wagons to Missouri, in order to bring back supplies for Maxwell and the ranch. That season there was more grass and water on the mountain branch of the Santa Fe Trail, and Kit decided to pass Bent's Fort and avoid the Cimarron Desert. He had gone only fifteen miles beyond the fork in the trail (the Cimarron Crossing of the Arkansas), when his little party ran into the Cheyenne village. Kit passed on, never dreaming of danger. Twenty miles farther on, a Cheyenne war-party visited his camp. One, two, three, five, seven, they drifted in as though by chance, until he found a score of warriors among his wagons.

Having always been on good terms with the Cheyennes, and being unaware that they were now at war with the whites, Kit invited the Indians to sit and smoke with him. They did so, talking among themselves in Sioux, so that the white men would not know the tribe to which they belonged. This made Kit curious, as he could see they were Cheyennes. Presently, in their talk, they dropped back into their own language. Kit had not listened to Making-Out-the-Road those two years for nothing. He could understand everything they said. And he realized that they had not recognized him at all. Yet he knew Stone Calf, the chief, quite readily. It must be the change in his appearance: he had bought some fofurraw fixens and got a haircut in St. Louis.

While Kit watched, he heard Stone Calf say, 'My friends, this white man has only fifteen men with him, and all but two of them are Mexicans. When he takes the pipe, he will lay aside his rifle. Then we can jump on him and knife him. After that, we can kill the Mexicans as easily as so many buffalo.'

When Kit heard that, he was alarmed. He quite

agreed with Stone Calf about the thirteen Mexicans. Hardy and enduring as the Spanish-Americans were, they were not rated as great fighters, especially when at a disadvantage. Kit knew the value of surprise. He jumped up, rifle in hand, the muzzle pointed toward the brown belly of the naked chief.

Said Kit: 'I know you. You are all Cheyenne warriors. I never done you any harm. I let you come into my camp as friends, and I have smoked with you. I don't know why you are talking of taking my hair. But I heard you. Now you get out. I will shoot any one who does not make tracks right now. And if you come back, I will fire on you.'

Stone Calf was furious, and ready to scalp the first white man he met. For an army officer, whose camp he had visited, had been drunk or crazy enough to flog the chief, in order to 'teach the redskins a lesson.' Kit was ignorant of this, and only his knowledge of the Cheyenne tongue and his quick courage saved him. The Indians left, but Kit knew very well that the incident was not closed. Warriors thronged the hill-tops.

'Catch up!' yelled Kit. 'Put out!' Slowly the wagons stretched out, and the frightened teamsters walked with a whip in one hand and a rifle in the other.

Kit corralled his wagons at dark, and had a talk soon after with his bravest and best Mexican. When Kit had finished, the young man nodded assent to Carson's proposal: '*Si, señor.*' Then he mounted Kit's own horse, and slipped out of the camp to ride and ride and ride those two hundred miles to the Rayado for help. Cautiously, he drifted out of the corral, out of the firelight, away into the emptiness of the prairie, lying flat on the horse's back, so that — if seen — the Indians might take him for a buffalo or a stray mule. After a mile or two, he sat upright, put spurs to his animal, and the lone horseman was off.

Kit kept strict watch that night, especially toward dawn, when Indians were most likely to attack. The teamsters dared not tether their animals outside the wagon corral, and so they had to cut grass with their skinning-knives for the cattle to eat. But at last the sun rose and found them already on their way. Kit made all the speed he could. He hoped to make Bent's Old Fort within a few days. At noon, when he halted, the Mexican must be well on his way, but Kit knew that it would be three or four days — perhaps a week — before help could arrive. And he knew that the headstrong Cheyennes would never wait that long to have their revenge.

Sure enough, at noon five Cheyennes rode into view. Kit had told them he would fire upon them, if they approached. But he did not, knowing that such a course would start the very fight he wished to avoid. Instead, he called them in to talk.

The five sat their horses opposite Kit, their leader — and Kit — each holding his open hand shoulder high in token of truce, while the talk went on. Kit said: 'Last night I sent a man on a fast horse to the soldiers on the Rayado. I have plenty friends in the white soldiers' camp. They are coming to help me fight. If you attack me, and I am killed, they will know that Stone Calf and his Cheyennes took my hair. They will never sleep until they rub you out. Look at me! I am Little Chief, who was hunter for Bent's big lodge on the Arkansas. You all know me. If there is fighting, I will not be the first to go under. *No-het-to!*'

Stone Calf said he would go and look for sign and see whether the white man spoke with a straight tongue or not.

The Cheyennes rode away, back toward Kit's camp of the night before. They must have found sign, for the wagon-train saw no more of them. The lone horseman had sixteen hours' start, and the Cheyennes knew they could not hope to overtake him. They went elsewhere to look for a victim.

Such a combination of courage, coolness, foresight, and intelligence made Kit the unique figure he was among the frontiersmen of his day. But he was not interested in his own cleverness and courage. What impressed him about this affair was the stupidity of the army officer who had flogged the chief of a friendly tribe because of some petty theft of a trinket in the soldier camp. Would a man declare war because the ambassador's wife stole his spoons at a banquet? Apparently, the army captain thought so. Kit was enraged at the folly of the men who put his life in danger — and all to no purpose.

Old Gabe Bridger was sufficiently impressed by the book l'arnin' of the army officers. He camped beside the Oregon Trail, halted every train until he found a set of Shakespeare, and then traded a span of oxen for the books — the best thar war, he had been told. Then he hired a boy at forty dollars a month to read it to him. One would give a good deal to have heard Old Gabe's opinion of Shakespeare. But Kit Carson on this occasion was not impressed by the intellectual attainments of the army officers.

For not only did he come near losing his life because of this officer's folly, but when his messenger reached Colonel Summer and asked for protection, the Colonel, whose own subordinate had done the flogging, refused to send aid. And so the messenger had to ride on until he found Major Grier. Grier at once rode at speed to rescue Kit, and, passing Summer's camp, made that officer understand the importance of his mission. Then the Colonel, to save his face, sent along Major Carleton and a detail. But Kit never gave him any credit for that. That was the army on the plains all over: first raise hell, and then neglect to take the necessary steps to redeem the error. No wonder Kit and Bent, and all the men who knew the Old West regarded the army as an unmixed curse to the country.

The troops found Kit — already saved by his own talents — not twenty miles from Bent's Old Fort, and the Indians were nowhere to be seen. Kit was warmly appreciative of Major Grier's prompt action, and never tired of praising that gallant officer's generous conduct. From the fort, Kit turned southward toward the Rayado, and reached the ranch in safety with all his wagons.

That was the last time Kit saw Bent's Old Fort, which he had helped to build. William Bent, now known to the Indians as Graybeard, was the sole survivor of the brothers, and had been trying to sell the fort to the War Department. The Indian trade was not what it had been. All the other old trading posts had been turned into garrisons: Bent's was the last to hold out. During the Mexican War, Bent had freely turned over his fort to the Government, to be used as a supply dépôt and army hospital, and for this he had never asked payment. He had also acted as guide and scout for the army, without other reward than the honorary title of colonel. And now that the War Department wanted the fort, he was asking less than it would cost to build it. But the waffle-tails in Washington would not pay Bent's price, and William Bent was not the man to haggle, or beg, or take a slap in the face. For thirty years he had called the fierce Cheyennes 'my people.'

One morning he ordered his men to load all he possessed upon his wagons and started them downriver. That night he camped on Wild Horse Creek, where Kit had fought the Crows just twenty years before. Next day he rode back alone to the Old Fort, walked through the great gloomy echoing *portal* into the silent sun-washed *patio*. He had ruled there like a king for twenty years, ruled an empire richer and larger than any conquered by Napoleon or Cæsar with all their legions, ruled it by sheer force of personality, swaying the councils of all the tribes from Mexico to the Black Hills. He set fire to the fort. In the bastions were kegs of powder. When the flames reached these, the old fort heaved itself into the air with a thunderous roar, and fell back a heap of ruins.

So, fittingly and in a moment, vanished the most historic building on the Old Santa Fe Trail. Where its graveled courtyard echoed to the hooves of Carson's hunter is now only a shallow depression on a bench by the lonely Arkansas, a depression surrounded by low gray heaps of crumbled adobes strewn with melted glass and hand-wrought iron nails, rusty as the memories of the days gone by.

The blowing up of Bent's Old Fort marked the end of the Old West, the heroic age of the last American frontier. After that came the Wild West, the age of the clowns and gunmen, the bandits and the land-grabbers, the gold-seekers and the railroad Hell-on-Wheels. Bent and Kit and the other men of the earlier time spent the rest of their lives saving what was left of the old order, security, and peace. But it was no use. The plains and mountains swarmed with unseasoned, untrained, undisciplined white men, ignorant, stupid, reckless, sometimes criminal. Nothing could prevent Indian wars under such conditions, and the coming of the army at last drove even the loyal Cheyennes into that career of bloodshed and rapine which made their name a terror to the border.

That same spring, Kit felt that he must have a taste of the good old days before it was too late. He could afford a vacation. Of course, that meant a trapping expedition. Not many of his old band were left. Death, California gold, army scouting, service as guides — all had taken toll. He and Maxwell could round up only eighteen of their old comrades of the beaver trail. With them Kit rode back into the mountains, where he had not set a trap for ten years. It was like the old thing to be once more among the fur.

They trapped the old Bayou, the South Platte, the Laramie Plains, the Parks, back to the Arkansas again. The beaver were plenty now, for no one in years had found it worth his while to disturb them, and that remnant of the Carson Men grained a many skin. Even the Indians obliged with a little fracas, to make the dramatized memory a complete success. But the details of the expedition will never be known.

They rode back over the Raton Pass to Taos as of old, half aware that the dust of their passing was the curtain to the drama of one of the most heroic of fighting brotherhoods.

The Carson Men would ride no more forever!

VII
INDIAN AGENT

CHAPTER XXI
FATHER KIT'S HAT

Soon after, Kit Carson had a chance to go to sea again — an opportunity which he was not slow to reject. Too well he remembered the voyage of the Navy Battalion of Mounted Riflemen! Maxwell and he had arrived in California, each with some six or seven thousand sheep, which they drove overland from New Mexico. They made an enormous profit on this expedition, successfully passing all the dangers of wolves, weather, and Indians, and Maxwell proposed a trip from San Francisco to Los Angeles. But Kit stuck to his mule and let Maxwell ride the wild waves.

Following the old trail by the Gila, Kit reached home on his birthday, Christmas, 1853. He found he had been appointed Indian agent. During the following spring, the Jicarilla Apaches went on the war-path and fought a pitched battle with the troops at Embudo on the road to Santa Fe. Losses were heavy on both sides, and soon after Lieutenant-Colonel Cooke set out to chastise the Indians, taking Kit along as guide.

The crossing of the Rio Grande del Norte was the first difficulty. The stream was in flood, and the crossing had to be made in the canyon, near the point where Arroyo Hondo enters the river. The bed there is full of rocks and very uneven, with many deep holes. The mules had to climb up and jump off the rocks repeatedly, and this made everybody a great deal of trouble. Kit led the way, and in time got them all over. A number of men narrowly escaped drowning. Then there was the canyon wall, hundreds of feet high, to climb. But at last Kit got them all up. During the day, in recrossing the horses in order to carry the infantry, Kit forded and swam that angry water some score of times. Within a few days the command surprised the Indians, killed several, and took their camp.

After the fight the Indians hurried away, and next day the troops pushed on after them, over snow and mountains, through canyons and water, rocks and sand. But the Apaches, scattering like so many quail, could not be run to earth. They led the troops in circles, so that sometimes the soldiers camped almost in the same place on consecutive nights. It was a game to the Indians, who, having no baggage and no animals, could keep out of the way of the troops, who must carry their rations with them. When mounted, the Apache's transport was also his commissary. He rode when speed was needed, and ate his horse when hunger urged. He always knew where to steal another animal.

While this fruitless chase was going on, a separate detachment came up with a Ute Indian, captured and disarmed him, and marched on. The Utes were at peace with the white men, and the captive Indian was afraid. Naturally, at the first opportunity, he made his escape. When Kit heard that story, and saw the weapons the men had taken from the Indian, he made haste to get in touch with the Ute chiefs, asking them to come and talk with him in Taos. Then he rode hard for home. The Apaches were bad enough; he wanted no trouble with his other charges, the Utes.

Soon after, the chiefs came in, and stood under the long *portal* of the Carson house. Kit beckoned them in, and the old men walked gravely through the thick-walled doorway, and sat upon the blankets stacked round the walls. With due observance of Indian custom, he offered them water, supposing that they must be thirsty after their ride. Then he offered them food, supposing they must be hungry. When all had eaten and drunk, he got out the long red-stone Cheyenne pipe he used for such councils, and, lighting it, passed it round. Then the talk began.

Father Kit, as the Utes called him, explained that the soldiers had captured their tribesman through a mistake, thinking him an Apache. The white men did not wish to fight with the Utes. He urged them to forget the matter, and said they must not help the Apaches in this war. For, if they did, the white men would fight them also.

Father Kit was a great man among the Indians, who respected him for his wisdom as well as for his imposing string of scalps. He never lied to them, and they knew it. The chiefs conferred, and said they would take his advice. Thereupon Kit returned the property taken from the Ute by the soldiers, made a few presents to the chiefs who had come so far, and they rode away.

Such prompt action inspired by real understanding of the Indians was Father Kit's chief merit as an Indian agent. He was not much good at reports, which must have been prepared by some friend or subordinate who could read and write. How Kit kept the accounts of his office is a mystery. But when it came to handling Indians, no other man in the West was his equal.

General Sherman bears witness to this, somewhat ruefully, as it was rather a shock to his personal vanity: the passage occurs in Rusling's 'Across America.' Sherman is quoted thus: 'These Red Skins think Kit twice as big a man as me. Why, his integrity is simply perfect. They know it, and they would believe him and trust him any day before me.'

Cooke having failed to conquer the Jicarillas, Major Brooks put out on their trail, but could not find them. Major Carleton then set his command in motion, with Kit as principal guide. Knowing the Indians well, Kit thought they would be making for the Mosco Pass in the White Mountains. When the troops arrived, they found the trail, as Kit had expected. This impressed Carleton, and he gave Kit a pretty free hand in directing the march. Kit and the guides kept the trail, while the soldiers followed through the timber, keeping back out of sight as much as possible, so as not to alarm the Indians. In this way, progress was made.

One morning, finding the trail fresh, Kit told the Major that he thought the Indians were not far ahead.

'We ought to find the Injuns about two o'clock,' said Kit.

Major Carleton was ready to give Kit credit for a good deal, but this sort of exact prediction, based upon the sign of a trail which the Major could not even see, was a little too much.

'Bet you a hat we don't!'

'Done,' said Kit. True mountain man, he never refused a wager.

On Fisher's Peak in the Raton Mountains, Kit found the Jicarillas — and the Major, getting out his watch, found the hour precisely *two*. After that, the Major listened to Kit. The campaign was a great success, and the two men remained fast friends. It was the first time Kit had seen an officer who could work with him and not bungle matters.

Having punished the Indians, Kit came back to Taos. Later, he made his way alone past the village of the hostile Apaches to the Ute village to ask the Utes to come to council. Such journeys were no unusual part of his work. He would have preferred to live in the camp with his Indian charges, where he could solve their difficulties immediately, and at the same time keep them from contact with the settlements, where they always came to grief one way or another. But the Indian Office — and Josefa — had other plans.

Some Mexicans had murdered a Ute and the Utes demanded payment. But the Indian agent could not make such payments. Kit promised to have the criminal arrested. But when this was done, nothing came of it, for the man escaped. This annoyed the Utes, who thought the whites were not guarding their interests.

On the way home they had the smallpox. The chiefs had been given new blanket coats, and every man who had a coat died. The Utes thought this was a plot against them and joined the raiding Apaches. Against such a combination the regular troops in that region were inadequate. Kit merely voiced the popular opinion when he said that, if the Indians were left to the citizens to handle, the wars would come to a sudden ending. The mountain men knew how to 'bring' the redskins.

Volunteer companies were organized, with St. Vrain at their head. They made a successful campaign. Seven battles were fought: seven victories were won by St. Vrain and his mountain men. Then the Utes and Apaches thought they would like to rest for a while, and against the judgment of all the citizens, including Kit Carson, a treaty was made. The whole thing would have to be done over again — nobody knew how soon.

A detailed account of these fights would be monotonous, and Kit was not mixed up in all of them. The chief result of all this campaigning was that Kit Carson got a new hat, a superb hat — of genuwine beaver!

This came from the States, and inside it was the inscription:

<u>At 2 o'Clock</u>

KIT CARSON

from

MAJOR CARLETON

VIII
PATRIOT

CHAPTER XXII
THE CIVIL WAR

WHEN Fort Sumter was fired upon, Kit Carson was in no doubt as to where he stood. He had not helped conquer the Indians and the Spaniards for nothing. Having added a territory larger and richer than the whole of the seceding States, he was not in the mood to join in the attempt to dismember the Union. Loyalty was one of his strongest traits. He was never half-hearted about anything.

Politics never interested Kit, and had never brought him anything but grief whenever they touched him. He had risked his skin a thousand times for the flag, and he never even considered the arguments brought forward pro and con. Moreover, Frémont was Commander of the Department of the West, with the rank of Major-General. Kit and his comrades in Taos at once took their stand for Abraham Lincoln and the Union.

The Confederates held Texas, of course, and soon had Arizona in their power. But that was all. Colorado, New Mexico, stood fast, and at once volunteers were organized. Carson was Lieutenant-Colonel of the First New Mexican Volunteer Infantry, with Ceran St. Vrain as Colonel. And after some months of preparation, the regiment took part in the battle of Valverde, February 21, 1862.

The fight was for the possession of the ford, and the Union forces were defeated. It was Kit's first experience of professional soldiers fighting professional soldiers, and he for some time supposed that his side was winning. As the West-Pointers had almost always gone about Indian fighting in the wrong way, not understanding it, so here Kit entirely mistook the nature of the problem. This was of no moment, however, for he had only to obey orders. During most of the time he was held in reserve. Kit was mentioned in dispatches. His regiment lost one killed, eleven missing, one wounded.

His part in the battle may be gathered from his own brief official report: '... in the afternoon I received ... the order to cross the river, which I immediately did, after which I was ordered to form my command on the right of our line and to advance as skirmishers towards the hills. After advancing some 400 yards we discovered a large body of the enemy charging diagonally across our front, evidently with the intention of capturing the 24-pounder gun, which, stationed on our right, was advancing and doing much harm to the enemy. As the head of the enemy's column came within some eighty yards of my right a volley from the whole column was poured into them, and the firing being kept up caused them to break in every direction. ... They did not attempt to reform, and the column, supported by the gun on the right, was moving forward to sweep the wood near the hills, when I received the order to retreat and recross the river. ... The column, after crossing the river, returned to its station near Fort Craig. ...'

The losses on both sides were negligible, considering

the numbers involved. As usual in those early battles of the Civil War, the recruits suffered more from excitement and nervous exhaustion than from anything else. Napoleon somewhere defined a battle as 'two bodies of men trying to frighten each other.' Where such bodies of men are made up of rookies, both sides are eminently successful in the attempt.

Kit Carson remained in service, but his duties during the years 1862–64 were those of the Indian fighter, and he had little concern with the white enemies of the Republic.

His first campaign was against the Apaches. For of course the Apaches took advantage of the distraction caused by the Confederates. They began to murder and rob and torture as usual. Carleton called upon Kit to teach them a lesson. Troops under his direction, though not directly under his command, soon caught the Apaches, and the chiefs came in and surrendered to Kit, whom they trusted.

Immediately Kit put in for leave — to see Josefa.

By all accounts his requests for leave were more frequent than any other letters sent by him through military channels.

Probably there never was a less soldierly officer in the armies of the United States than Kit Carson. He had been so long accustomed to leading men who could take care of themselves that it was extremely difficult for him to mother the regulars; to look out for men who had to be told when to wear their overcoats, when to clean their guns, when to bathe, shave, and perform the other functions of a human animal. Kit's superiors — well-trained but unindulgent grandmothers — had to be continually reminding him of the importance of latrines, kitchens, forage, rations: Kit could never bring himself to count socks or shoe-strings, never got used to men who would throw their equipment away to avoid carrying a heavy pack, men who — if not prevented — would eat up all their week's rations at one sitting. His reaction to the army was that of the Indian in the late war, who complained, 'No good: too much salute; not much shoot.'

Rank and authority meant little to Colonel — or General — Carson. He always insisted that the boys call him Kit. And as for the traditions of the Service!!! He had no dignity to maintain, and used to lie on a blanket in front of his quarters, romping with his children by the hour.[1] And he never expected any better treatment than his men received. They admired him, loved him, and took advantage of him in the matter of written passes, offering him papers to sign which gave them privileges of which he never dreamed.[2] For by this time Kit had learned to scratch his name, though he had no idea of the letters which composed it. It was a Chinese pictograph to Kit.

In the field he could give the martinets cards and spades and take the pot. He remained half-Indian, half-mountaineer all his life. Naïvely, he supposed the business of a soldier was to fight!

Next in order was a campaign against the Navajo of the desert. The Navajo were numerous, hardy, brave, and had been uniformly successful for centuries against the Mexicans, whom they cordially hated. No tribe west of the plains was so self-sufficient. The blood of the tribe contained a large measure of Spanish and American corpuscles, and no people were better

[1] Sabin, *Kit Carson Days*, p. 398.   [2] *Ibid.*, pp. 425–26.

able to fight, or give plausible reasons for their fighting. When they found their country included in the territories of the United States, following the Mexican War, they still wished to go on fighting the Mexicans, and complained to the American General, saying, 'This is *our* war!'

After 1847, three campaigns had been launched against them by the armies of the United States, and all three were failures. This was not merely due to the courage and address of the Indians and the inexperience of the commanding officers, but to the almost impassable nature of the Navajo country, which — even to this day — has not been traveled much by white men.

Desert, and mountain, and gorge, the Navajo had an admirable region for defense. Apache land was bad enough. But the canyon land of the Navajo was worse. And of all the canyons, the famous Canyon de Chelly was worst and most inaccessible. Thither the tribesmen always retreated when hard pressed, safe among their caves and stone forts, their peach orchards and their flocks of sheep, their cornfields and their melon patches. Treaties were made and broken the same day, and until Kit Carson took charge, nothing was accomplished. Some of the credit must be given Major Carleton, whose subordinate Carson was, though the Major himself did not take the field.

In January, 1864, having mopped up all the outlying country and driven the Navajo to take refuge in their canyon, Kit set out with enough men to invade even this. He himself, being in charge, took no part in the actual fighting. But for all that, the conquest of the Navajo is unquestionably one of his greatest achievements.

The Canyon de Chelly, about thirty miles long and with sheer rock walls towering hundreds — perhaps a thousand feet — in the air, with an icy stream at the bottom, and innumerable niches and coigns of vantage where the prehistoric cliff-dwellers had built their nests, was no easy nut to crack. But Kit Carson for once had all the men he needed. He sent one column through the bottom, others along the rims, and cleaned it all out from end to end. The result of his efforts was an unconditional surrender, and the concentration of some seven thousand Navajo prisoners. The Gibraltar of the Navajo was taken, and from that day they have been quiet, thrifty, law-abiding citizens, increasing in number and wealth from year to year.

These two successful expeditions made the country realize that Kit Carson was the greatest Indian fighter in the army. He had crushed the Apaches, the most dreaded and crafty of mountain Indians. He had tamed the Navajo, most manly and treacherous and numerous of all the tribes of the desert. Only one world remained for him to conquer: the hostiles of the Southern Plains.

IX
PEACEMAKER

CHAPTER XXIII
ADOBE WALLS

SOME years before 1840, Bent and St. Vrain built a large trading fort in the range of the Kiowas and Comanches, who dared not venture north to trade at Bent's Old Fort on the Arkansas. After 1840, when these tribes had made peace with the Cheyennes and Arapahoes, all four peoples traded at the Old Fort, and the southern establishment was abandoned and allowed to fall into decay.

This ruin was known as Adobe Walls. It stood on the Canadian River, in the Panhandle of Texas. Nothing remains of it now. But the name is familiar to every lover of the Old West. For on that ground took place two fights: one the biggest Indian battle ever fought on the plains; the other the most celebrated skirmish with Plains Indians now on record.

The first was Kit Carson's battle of Adobe Walls, November, 1864, when he attacked the combined camps of the allied tribes. The second was the little affair of the buffalo hunters, when Billy Dixon and his comrades stood off the enraged Cheyennes and Comanches, whose treaty rights they were violating.

Billy Dixon's well-written 'Life' has given his involuntary defense of Hanrahan's Saloon a greater fame than is enjoyed by Kit Carson's pitched battle on the same ground. Kit's memoirs came to an end before that time, and most of his biographers are a bit vague about his biggest and best enterprise. Fortunately, Captain George H. Pettis, who served with Kit on the Canadian, has left us a detailed record of what he saw. All later accounts must largely follow his, although supplemented by the statements of the Indians concerned.

Kit was placed in command of some three hundred men, one hundred of whom were infantry. He had two mountain howitzers, twenty-seven wagons, an ambulance, and some threescore Ute and Jicarilla Apache scouts. He had rations for forty-five days, and was given a free hand. Carleton gave Kit all and more than he asked for, told him where to look for the Indians, and said, in effect, 'Go to it!'

Here was Kit Carson's grand opportunity to show what thirty-eight years of Indian fighting, high courage, and clear understanding of the country and the problems before him could do.

Kit's command marched downstream along the Canadian, heading for Adobe Walls, where Kit intended to make his headquarters. From that point he would ride out with cavalry and pack-trains to lick the hindsights off the cussed Injuns. But, as it happened, the pack-mules were unnecessary. For the Indians were waiting for him at Adobe Walls.

Kit knew the country well. He had been on the ground often, as one of Bent's men, and we have Grinnell's account of a little scrape with Kiowas and Comanches just there, when Kit lost his animals and had to *cache* Bent's packs and walk home through the cactus, along with Fisher, Murray, and Blackfoot Smith.[1] That was twenty-five years back.

[1] Grinnell, *The Fighting Cheyennes*, chap. XXIV.

Passing the spot where the merchant White had been murdered, Kit led his men on, throwing out scouts on flank and in the van, and at night sleeping calmly through all the hideous racket of the Indian war dance kept up by the Utes and Jicarillas. They knew that Father Kit would bring them to the enemy, and with Indian canniness, hardened their muscles and raised their morale by dancing and whooping and rehearsing the coming fight every night of the march.

'We arrived at Mule Spring early in the afternoon; had performed our usual camp duties, and as the sun was about setting, many of us being at supper, we were surprised to see our Indians, who were lying around the camp, some gambling, some sleeping, and others waiting for something to eat from the soldiers' mess, spring to their feet, as if one man, and gaze intently to the eastward, talking in their own language.... Colonel Carson ... informed us that the two scouts that he had dispatched that morning, had found the Comanches, and were now returning to report the particulars. Although the returning scouts were at least two miles distant, and, mounted on their ponies, were hardly discernible, yet the quick, sharp eye of our Indians made them out without difficulty.... And what was more remarkable, they had ... conveyed the intelligence that they had found the enemy, and that there was work to be done.... They reported that ... we should have no difficulty in finding all the Indians that we desired.'[1]

That night Kit marched fifteen miles or so on the trail of the hostiles, and kept his men ready to mount

[1] *Personal Narratives of the Battles of the Rebellion*, No. 5. Providence, Rhode Island, 1878.

all night. At dawn, Kit told of a dream he had had the night before, a dream of battle and howitzers firing. Kit believed in dreams, and sure enough, before he had completed his story, cries were heard across the river, which told Kit that he had found his enemy.

In a moment, the Utes and Jicarillas discarded their buffalo robes, and charged naked into the river after the foe, eager to get as many of the Comanche horses as possible. Kit could see the ponyherds racing downriver to their camp, and ordered his cavalry and cannon forward, while the triumphant Utes came galloping back, each scout having as many captured horses as he could wrangle. Remounting, the Indians joined the attack once more.

Kit, true to his Indian training, felt like stripping for action now, and peeled off his heavy overcoat and threw it into the bushes, although the frost was heavy on the grass. The cavalry in front was firing, and Kit could tell from the rapid recession of the sound that the Indians were giving way. About nine o'clock, Kit could see the trim white tepees strung thickly along the river, five miles in advance. The army officers naïvely supposed that the tepees were Sibley tents, and that they had run into General Blunt's column. But Kit knew better, and, galloping forward, ordered Pettis to 'throw a few shell into that crowd over thar.' For the Indians, having run, were now coming back to make a stand and cover the retreat of their women and children. The troopers had dismounted, corralled their horses safely within the thick walls of the old fort, and were deploying as skirmishers.

The battery went into action on a small knob in the middle of the flat, grassy bottoms, and on this little elevation, Kit took his stand to direct operations. The cavalrymen now lay down in the tall grass, and fired occasionally, while the Utes and Jicarillas were staging a lively exhibition of horsemanship and bravado in front of the line of some two hundred mounted hostiles. These charged, 'in the same manner, with their bodies thrown over the sides of their horses, at a full run, and shooting occasionally under their horses' necks; while gathered just beyond them twelve or fourteen hundred, with a dozen or more chiefs riding up and down their line haranguing them, seemed to be preparing for a desperate charge on our forces. Surgeon Courtright had prepared a corner of Adobe Walls for a hospital, and was busy, with his assistants, in attending to the wants of half a dozen or more wounded. Fortunately, the Adobe Walls were high enough to protect all our horses from the enemy's rifles, and afford ample protection to our wounded.'

Kit had now marched clear through the first village — that of the Kiowa-Apache — but had not had time to stop and burn it. He had to drive away the Indians in his front first. When the battery began to fire, the hostiles 'rose high in their stirrups and gazed, for a single moment, with astonishment, then guiding their horses' heads away from us, and giving one concerted, prolonged yell, they started on a dead run for their village. In fact when the fourth shot was fired there was not a single enemy within the extreme range of the howitzers.'

Kit thought breakfast was next in order, and his men unsaddled, watered, tethered the horses to graze, and began to prepare their food. After breakfast, Colonel Carson intended to march down-river and burn the other camps, or, at any rate, destroy the one he had captured.

While the men were congratulating themselves upon the victory, Kit reconnoitered. Through his glass he discovered a large force of Indians advancing from another village about three miles down-river from Adobe Walls. That village, he estimated, contained about three hundred and fifty lodges. At once Kit ordered his men to saddle, mount, and form. Soon after he found himself besieged by 'at least a thousand warriors mounted upon first-class ponies. They repeatedly charged my command from different points,' says his official report.

The fight went on all afternoon, and the Kiowas and Comanches soon got used to the cannon and learned to avoid the shells. One Indian was hit by a shell, the missile passing through the body of his horse, which was running at top speed. The Comanche was thrown, and lay senseless, but two of his comrades performed the rescue, carrying him off between them at the gallop, while the whites poured bullets after them.

Pettis goes on to tell how the Indians dismounted and took cover in the tall grass, using their excellent marksmanship to great advantage, while yet others continually charged on horseback. In rear of their lines some one kept sounding off on a bugle, and all day long he would blow the opposite call to that used by Carson's buglers. He kept it up all during the fight, to the amusement of the volunteers. Kit thought it must be a white man, but the Indians tell a different story. One of their chiefs had a bugle, which he used expertly, probably with the idea that it was big soldier medicine!

Do-hau-sen led the Kiowas, and Stumbling Bear distinguished himself in the fight, counting not less than three *coups*. Ironshirt, the Kiowa-Apache chief, was shot at the door of his tepee.

And now came more Comanches, more Kiowas, some Arapahoes, all as well armed as the soldiers, and better mounted. Not less than three thousand were out to get Kit's hair. They had guns, powder, and ball in abundance, countless fine horses, and they outnumbered the soldiers ten to one. Surrounding the whites, they rode back to their captured village and carried away their property under Kit's nose. And back there, coming slowly along, unconscious of any danger, was Colonel Carson's train of twenty-seven wagons and one ambulance. If the Indians found that, the booty would be all theirs before a man could smoke a pipe.

Pettis and the other officers urged Kit to attack the villages downstream. They thought things were going splendidly. But Kit, and the Utes and the Jicarillas, never dreamed of such an enterprise. They knew all too well what would become of them if such a foolhardy attempt were made. It made the Utes frantic to see the Kiowas run off the ponies they had captured earlier. In the middle of the afternoon, Kit formed his cavalry led horses in column of fours, deploying his dismounted men in the rear and on the flanks.

'In this manner,' says Kit, 'I commenced my march on the village.' 'The village' was the one he had previously taken, not the one ahead. At once the Indians, seeing what he was up to, charged furiously and again and again, until it looked as though Kit's little force were done for. But the cavalrymen stood their ground, fired steadily, and 'caused them to retire ... with great slaughter.'

That was bad enough. But presently Kit saw the Indians setting fire to the tall, dry grass of the river bottoms — grass taller than a man on horseback! The wind was right, and in a moment the flames were sweeping up the valley on the heels of his soldiers. It was the old Indian trick. Kit ordered the grass ahead of him to be fired at once, so as to clear his path, and then led his men and frightened horses up the slopes to a hillside where the grass was thin and short. The flames swept on, and behind the smokescreen came the Indians, riding their fast horses, armed with carbines, eager to strike a *coup* from the cover of the smoke. Again and again they charged up; the white men had never seen redskins exhibit such bravery before. It was hot work.

One scalp, and one only, was taken that day from an Indian head. A young Mexican recruit, who had been bitten that morning by a rattlesnake, was so desperate that he fought recklessly. Once, when the wind suddenly parted the curtain of smoke, he caught sight of a Comanche swooping past. *Crack!* Down went the redskin, and out through the smoke ran the Mexican, knife in hand, and stripped off the trophy before the Indians could run up to rescue their fallen comrade. He came back yelling and waving the hair. But in all Kit's command, no one else had either the time or the inclination to scout outside the lines for Indian hair that day.

Kit never saw his overcoat again, and had to ride home through the frosty weather wrapped in a captured buffalo robe, like one of his own scouts.

Near sundown the command reached the Kiowa-Apache village, which they had captured early in the morning. A few shells frightened away the Indians who had come back to get their property. The soldiers charged in, looted the camp, and burned the lodges. They found enough fine robes to supply every man in the command. Also they found clothing of white women and children, United States Army hats and sabers, and a remodeled ambulance in which Little Mountain (or Do-hau-sen), the Kiowa chief, was accustomed to ride in state, with Indian boys riding the half-broken ponies. All these things they burned with the lodges.

The detail which fired the tents found the women of the Ute scouts rejoicing, and proudly displaying the bodies of four Kiowas — two crippled, and two blind — whom they had found and killed with an axe. The soldiers reported this to Kit, but it was too late.

Of the Kiowas murdered by the Utes, some were women. So much has been imputed to the frontiersmen in the matter of killing Indian women, that it is worth while to quote Kit's own opinion of such action, an opinion which he expressed in 1866 after hearing of the terrible massacre of friendly Cheyennes by Colonel Chivington, which occurred three days after his own battle at Adobe Walls. The passage is found in General Rusling's 'Across America':

After telling how he had once seen an Indian kill his own brother for insulting a white man in the old times, Kit said the Indians, in their outrages, were only imitating or improving on the white men's bad example.

Said he, 'To think of that dog Chivington, and his

hounds, up thar at Sand Creek! Whoever heerd of sich doings among Christians! The pore Injuns had our flag flyin' over 'em, that same old stars and stripes thar we all love and honor, and they'd been told down to Denver, that so long as they kept that flyin' they'd be safe. Well, then, here come along that durned Chivington and his cusses. They'd bin out several days huntin' hostile Injuns, and couldn't find none no whar, and if they had, they'd run from them, you bet! So they just pitched into these friendlies, and massa-*creed* them — yes, sir, literally massa-*creed* them — in cold blood, in spite of our flag thar — women and little children even! Why, Senator Foster told me with his own lips (and him and his committee investigated this, you know) that that thar damned miscreant and his men shot down squaws, and blew the brains out of little innocent children — even pistoled little babies in the arms of their dead mothers, and worse than this! And ye call *these* civilized men — Christians; and the Injuns savages, du ye?

'I tell ye what; I don't like a hostile Red Skin any better than you du. And when they are hostile, I've fit 'em — fout 'em — as hard as any man. But I never yit drew a bead on a squaw or papoose, and I loathe and hate the man who would. 'Taint nateral for brave men to kill women and little children and no one but a coward or a dog would do it. Of course, when we white men do sich awful things, why, these pore ignorant critters don't know no better, than to follow suit. Pore things! I've seen as much of 'em as any white man livin', and I can't help but pity 'em.'

Too late to save the blind squaws, Kit led his men away from the blazing village, and, marching through the darkness on his back trail, anxiously looked for his lost wagon-train.

The two gun carriages and the two ammunition carts served as temporary ambulances for the wounded, who suffered severely. After more than twenty-four hours of marching and fighting, both men and horses were exhausted. Nobody knew when the Indians might come charging out of the darkness, or when an arrow might be slipped between his ribs. All the supplies were with the wagon-train, and the nearest settlements more than two hundred miles away. It was a slow march, an anxious march, a retreat. But it was made in good order.

Even Kit's Utes were so tired that they omitted the scalp dance that night — over the one scalp the envenomed Mexican had taken. Oh, how good that challenge sounded to the tired troopers: 'Who comes there?' How the men slept! How the salt pork and hardtack and coffee vanished! Next day, Kit decided that 'it was impossible for me to chastise them further at present,' and set out for home, with thirty of his forty-five days' rations still unconsumed.

Short of extermination, no other victory could be won against Plains Indians. Run, and they follow: follow, and they run. Kit ended his brilliant attack by that most difficult of military feats — a retreat before superior numbers *after* a decisive defeat.

One might suppose that men of action — soldiers — would speak straightforwardly about their exploits — and defeats. But official *communiqués* are as cloudy and magniloquent as a speech on tariff reform by a Republican President. Whoever wrote Kit's report for him made the battle of Adobe Walls a glorious victory, and thereby robbed Kit of the credit of his real achievement.

For there is no doubt that the Indians had him licked. He never denied it, and often said that it was only the old fort and the howitzers which prevented his command from being wiped out that day. Every witness — except the official report — called that battle a defeat. And it was Kit's greatest triumph.

Nevertheless, it *was* a triumph, and deserves a fame which has been given to lesser men who were better advertisers. The most famous of these gallant Indian fighters was General Custer. On two occasions he tried to do what Kit succeeded in doing. It is worth while to compare the merits of these attempts.

Custer's two most important engagements were the battle of the Washita in 1868, and the *débâcle* on the Little Bighorn, in which he lost his life, 1876. Both were precisely similar attempts to that of Kit Carson at Adobe Walls. In both, Custer attacked a smaller camp, and in both he suffered greater losses than Kit, without inflicting any more damage.

The battle of the Washita was almost identical: an attack upon the same Indians, at dawn, with cavalry, resulting in a capture of the first of a long string of camps, some ponies, and a number of women and children. Owing to the fact that it was mid-winter and the Indians' ponies were no good, Custer made his getaway before the Kiowas and Cheyennes could get together and rub him out. Custer counted this a glorious victory. And he is commonly given the credit for the novel idea of attacking Indian camps in winter. All of these things Kit did first, and did better.

No one, I suppose, will claim that Kit could **have**

had worse luck than Custer had on the Little Bighorn. The fact is, Custer was in bad odor with his superiors, and in a desperate attempt to come back, risked everything when the cards were stacked against him. He tried to do to the Sioux camp what he had done to the Kiowas on the Washita. But it was then summer, and the Sioux ponies were in flesh. The result is known to every one. The Indians rubbed Custer out within half an hour after he sighted the village.

At Adobe Walls, Kit killed more Indians than Custer on the Washita. He fought them off all day, whereas Custer cleared out at once, leaving Major Elliot and his other dead unburied for two weeks. If *that* is a victory, Kit Carson's battle was a march of triumph. Unfortunately, Kit Carson was no hand at wearing fancy jackets and writing up his exploits. It never occurred to him that what was said about a deed was greater than the deed itself. Like Custer, he rode out to look for Injuns, found them, stood them off a while, and then very wisely put his tail between his legs and ran away.

No doubt, if Kit had failed and Custer succeeded, the battle of Adobe Walls would be as famous as the Little Bighorn is now. Spectacular failure is always more arresting than that unsatisfactory compromise commonly called success. But however the public may regard it, this fight of Kit's was certainly the biggest of all the battles fought in the West, if we measure them by the number of Indians engaged. And it was one of the most successful, and the crowning achievement of the wariest and boldest Indian fighter who ever lifted hair.

Nevertheless, it was a defeat.

Those Plains Injuns!

CHAPTER XXIV
THE LAST SMOKE

THE last years of Kit Carson were uneventful. He spent them guarding the Santa Fe Trail against raiding Indians, riding up and down the ranges and across the plains to help Uncle Sam make vain treaty after treaty with the restless tribes — Comanches, Kiowas, Cheyennes, Apaches, Arapahoes — who found themselves caught in the tightening ring of white settlements, saw their buffalo slaughtered, their lands invaded and criss-crossed by wagon trail and railroad, saw, in short, their finish, and could only fight like wild things in a trap. Kit stuck to Carleton and the army as long as he could, and made several trips to Washington on official business, in the intervals giving as much time as possible to his numerous children and Josefa.

But his health was going. While on a hunting trip, just before the Civil War, he had been thrown from his horse, dragged some distance, and badly hurt internally. The medical science of those days could not cope with his injuries, and at last in 1867 he resigned his commission and went home, a sick man, to settle down, as he thought, for good. Kit was much emaciated, the mere shadow of his former self, and he felt entitled to a rest.

But the Utes threatened trouble, and he was asked to take a delegation of chiefs to Washington as the one hope of avoiding war. As usual, public spirit got the better of his native caution: he went. He came back worn out, sicker than ever, to find Josefa on the verge of another confinement. They were living near the mouth of the River of Lost Souls on the Arkansas, along with a few other old-timers who clung together amid the changes sweeping their world away. There Josefa died in childbirth, leaving Kit broken-hearted, with a large family upon his failing hands.

The worst evils that can befall a man were heaped upon him in those last days: he was sick, he was broke, he was lonesome. Small wonder that the inevitable march of his disease quickened. He was so feeble that he left his responsibilities to relatives and friends, and moved to Old Fort Lyon, not far off, where he could be under Surgeon Tilton's eye. But the Surgeon could give him no hope.

At his request they made him a bed of buffalo robes on the floor in a house on Officers' Row. There he made arrangements for the disposition of his small estate: his house in Taos, his ranch, his bunch of Cross-J cattle, and requested that his bones be buried beside Josefa's, in Taos, his lifelong home. He suffered a great deal, but maintained his gentle, simple, courageous heart throughout the tedious days. His only diversion was talk or listening to some one read to him. He was on a light diet, and his one dissipation — his pipe — was forbidden him.

In those last days he had the reward of all his years of hardship and self-sacrifice: honor, love, obedience, troops of friends, who loved him for himself, as well as for the dangers he had passed. But Josefa was gone, and Kit had nothing left to live for. It seems to have been his expectation to live a hundred years, for he repeatedly complained that, if it had not been for his disease, he might have done so. After all he had been through, it was no wonder if he thought he bore a charmed life. It was May, 1868, and he was not yet sixty.

At the end of the week, he seemed brighter, and showed more spirit than he had previously. His steady blue eyes shone with some of their old fire. Mr. Scheurich, the godfather of his children, attended him, and from the presence of his *compadre* Kit appeared to derive great satisfaction. 'About mid-afternoon, of May 23, he said that he was hungry, and he asked Mr. Scheurich to cook him a good dinner; ... he was tired of what had been given him.'[1]

'Cook me some fust-rate doin's,' said Kit. 'A buffalo steak and a bowl of coffee and a pipe are what I need.'

The army surgeon warned Kit that the meal would probably be fatal. But Kit insisted, and the surgeon, knowing that he was going soon, did not long oppose him. 'Accordingly, Mr. Scheurich ... cooked a substantial steak and made coffee, and brought them in; and General Carson dined heartily. ... Then he called for a pipe. Which pipe would he have? There was the fine pipe, gift to him from General Frémont, and there were other pipes. ... No; he wished his old clay pipe, and while he smoked it he and his *compadre* would talk of old times. ... He smoked; Mr. Scheurich chatted; Surgeon Tilton listened. Suddenly the General coughed.'[2] The expected hemorrhage followed. The aneurism had ruptured into the trachea. Kit called out: 'I'm gone! Doctor, *compadre, adios!*' The end was swift.

[1] Sabin, *Kit Carson Days*, p. 496.   [2] *Ibid.*, p. 497.

General Christopher Carson was buried with military honors in the cemetery of that lonely post, and as the country then afforded no flowers for his coffin of rough planks, the ladies of the fort 'gave the white paper flowers from their bonnets.'[1]

Afterward, Kit and Josefa were laid side by side at Taos.

So died Kit Carson, brave, unaffected, self-sufficient to the last puff of his old dudheen, a valiant trencherman, with the bull meat under his belt, and the old gleam in his tired eyes, blowing smoke into the jaws of Death, whom he had flouted so often.

No more fitting end could have been desired to the drama of his life of wandering and fighting, had he been vain enough to posture and pose as the curtain went down.

But there was no pose in Kit Carson, and the West may hold his name high above the movie cowboys, the Wild West showmen, the gaudy, strutting soldiers, the cruel killers, who clamor down the old, loyal, patient courage of the pioneer. For Kit was greater than them all.

>This is the happy warrior; this is he
>That every man in arms should wish to be.

[1] Sabin, *Kit Carson Days*, p. 498

THE END